I0488641

Groundwater Quality in the Columbia Plateau, Snake River Plain, and Oahu Basaltic-Rock and Basin-Fill Aquifers in the Northwestern United States and Hawaii, 1992–2010

By Lonna M. Frans, Michael G. Rupert, Charles D. Hunt, Jr., and Kenneth D. Skinner

National Water-Quality Assessment Program

Scientific Investigations Report 2012–5123

U.S. Department of the Interior
U.S. Geological Survey

U.S. Department of the Interior
KEN SALAZAR, Secretary

U.S. Geological Survey
Marcia K. McNutt, Director

U.S. Geological Survey, Reston, Virginia: 2012

For more information on the USGS—the Federal source for science about the Earth, its natural and living resources, natural hazards, and the environment, visit http://www.usgs.gov or call 1–888–ASK–USGS.

For an overview of USGS information products, including maps, imagery, and publications, visit http://www.usgs.gov/pubprod

To order this and other USGS information products, visit http://store.usgs.gov

Suggested citation:
Frans, L.M., Rupert, M.G., Hunt, C.D., Jr., and Skinner, K.D., 2012, Groundwater quality in the Columbia Plateau, Snake River Plain, and Oahu basaltic-rock and basin-fill aquifers in the northwestern United States and Hawaii, 1992–2010: U.S. Geological Survey Scientific Investigations Report 2012–5123, 84 p.

Foreword

The U.S. Geological Survey (USGS) is committed to providing the Nation with reliable scientific information that helps to enhance and protect the overall quality of life and that facilitates effective management of water, biological, energy, and mineral resources (http://www.usgs.gov/). Information on the Nation's water resources is critical to ensuring long-term availability of water that is safe for drinking and recreation and is suitable for industry, irrigation, and fish and wildlife. Population growth and increasing demands for water make the availability of that water, measured in terms of quantity and quality, even more essential to the long-term sustainability of our communities and ecosystems.

The USGS implemented the National Water-Quality Assessment (NAWQA) Program in 1991 to support national, regional, State, and local information needs and decisions related to water-quality manage¬ment and policy (http://water.usgs.gov/nawqa). The NAWQA Program is designed to answer: What is the quality of our Nation's streams and groundwater? How are conditions changing over time? How do natural features and human activities affect the quality of streams and groundwater, and where are those effects most pronounced? By combining information on water chemistry, physical characteristics, stream habitat, and aquatic life, the NAWQA Program aims to provide science-based insights for current and emerging water issues and priorities. From 1991 to 2001, the NAWQA Program completed interdisciplinary assess¬ments and established a baseline understanding of water-quality conditions in 51 of the Nation's river basins and aquifers, referred to as Study Units (http://water.usgs.gov/nawqa/studies/study_units.html).

National and regional assessments are ongoing in the second decade (2001–2012) of the NAWQA Program as 42 of the 51 Study Units are selectively reassessed. These assessments extend the findings in the Study Units by determining water-quality status and trends at sites that have been consistently monitored for more than a decade, and filling critical gaps in characterizing the quality of surface water and groundwater. For example, increased emphasis has been placed on assessing the quality of source water and finished water associated with many of the Nation's largest community water systems. During the second decade, NAWQA is addressing five national priority topics that build an understanding of how natural features and human activities affect water quality, and establish links between sources of contaminants, the transport of those contaminants through the hydrologic system, and the potential effects of contaminants on humans and aquatic ecosystems. Included are studies on the fate of agricultural chemicals, effects of urbanization on stream ecosystems, bioaccumulation of mercury in stream ecosystems, effects of nutrient enrichment on aquatic ecosystems, and transport of contaminants to public-supply wells. In addition, national syntheses of information on pesticides, volatile organic compounds (VOCs), nutrients, trace elements, and aquatic ecology are continuing.

The USGS aims to disseminate credible, timely, and relevant science information to address practical and effective water-resource management and strategies that protect and restore water quality. We hope this NAWQA publication will provide you with insights and information to meet your needs, and will foster increased citizen awareness and involvement in the protection and restoration of our Nation's waters.

The USGS recognizes that a national assessment by a single program cannot address all water-resource issues of interest. External coordination at all levels is critical for cost-effective management, regulation, and conservation of our Nation's water resources. The NAWQA Program, therefore, depends on advice and information from other agencies—Federal, State, regional, interstate, Tribal, and local—as well as nongovernmental organizations, industry, academia, and other stakeholder groups. Your assistance and suggestions are greatly appreciated.

William H. Werkheiser
USGS Associate Director for Water

Contents

Contents—Continued

Figures

Figures—Continued

Figures—Continued

Tables

Conversion Factors, Datums, and Abbreviations and Acronyms

Conversion Factors

Multiply	By	To obtain
Length		
inch (in.)	2.54	centimeter (cm)
foot (ft)	0.3048	meter (m)
mile (mi)	1.609	kilometer (km)
Area		
acre	0.4047	hectare (ha)
square mile (mi^2)	259.0	hectare (ha)
Volume		
acre-foot (acre-ft)	1,233	cubic meter (m^3)
cubic foot (ft^3)	0.02832	cubic meter (m^3)

Temperature in degrees Fahrenheit (°F) may be converted to degrees Celsius (°C) as follows:

$$°C=(°F-32)/1.8.$$

Datums

Vertical coordinate information is referenced to the North American Vertical Datum of 1988 (NAVD 88).

Horizontal coordinate information is referenced to the North American Datum of 1983 (NAD 83).

Altitude, as used in this report, refers to distance above the vertical datum.

Abbreviations and Acronyms

CFC	Chlorofluorocarbons
DBCP	Dibromochloropropane
DCP	1,2-dichloropropane
EDB	1,2-Dibromoethane
GIS	Geographic Information System
HBSL	Health-Based Screening Levels
LRL	Laboratory reporting limit
LT-MDL	Long-term method detection limit
MCL	Maximum contaminant level
NAWQA	National Water-Quality Assessment Program
NWQL	National Water Quality Laboratory
PCE	Tetrachloroethene
ROC	Receiver Operating Characteristic
STATSGO	State Soil Geographic
SSURGO	Soil Survey Geographic
TCE	Trichloroethene
TCP	1,2,3-trichloropropane
VOC	Volatile organic compound

Groundwater Quality in the Columbia Plateau, Snake River Plain, and Oahu Basaltic-Rock and Basin-Fill Aquifers in the Northwestern United States and Hawaii, 1992–2010

By Lonna M. Frans, Michael G. Rupert, Charles D. Hunt, Jr., and Kenneth D. Skinner

Abstract

This assessment of groundwater-quality conditions of the Columbia Plateau, Snake River Plain, and Oahu for the period 1992–2010 is part of the U.S. Geological Survey's National Water Quality Assessment (NAWQA) program. It shows where, when, why, and how specific water-quality conditions occur in groundwater of the three study areas and yields science-based implications for assessing and managing the quality of these water resources. The primary aquifers in the Columbia Plateau, Snake River Plain, and Oahu are mostly composed of fractured basalt, which makes their hydrology and geochemistry similar. In spite of the hydrogeologic similarities, there are climatic differences that affect the agricultural practices overlying the aquifers, which in turn affect the groundwater quality. Understanding groundwater-quality conditions and the natural and human factors that control groundwater quality is important because of the implications to human health, the sustainability of rural agricultural economies, and the substantial costs associated with land and water management, conservation, and regulation.

The principal regional aquifers of the Columbia Plateau, Snake River Plain, and Oahu are highly vulnerable to contamination by chemicals applied at the land surface; essentially, they are as vulnerable as many shallow surficial aquifers elsewhere. The permeable and largely unconfined character of principal aquifers in the Columbia Plateau, Snake River Plain, and Oahu allow water and chemicals to infiltrate to the water table despite depths to water commonly in the hundreds of feet. The aquifers are essentially unconfined over large areas, having few extensive clay layers to impede infiltration through permeable volcanic rock and alluvial sediments. Agriculture is intensive in all three study areas, and heavy irrigation has imposed large artificial flows of irrigation recharge that rival or exceed natural recharge rates. Fertilizers and pesticides applied at land surface are leached from soil and transported to deep water tables with the infiltrating irrigation recharge, resulting in a layer of degraded water quality overlying better quality regional groundwater beneath. This "irrigation-recharge layer" is best known on Oahu, where

it has been studied since the 1960s; however, the extent of nitrate and pesticide contamination in the Columbia Plateau and Snake River Plain indicate that the same situation exists in those areas. Contamination from agricultural and urban activities is present not only at shallow depths in surficial materials of the three areas, but extends regionally in the deep, principal bedrock aquifers that are tapped for drinking water by domestic and public-supply wells.

Naturally occurring constituents and nitrate concentrations above human-health benchmarks—Maximum Contaminant Levels (MCLs), and Health-Based Screening Levels (HBSLs)—were more common in the Columbia Plateau and the Snake River Plain than in Oahu. Concentrations of anthropogenic constituents (constituents related to human activities) above human-health benchmarks were more common in Oahu. Naturally occurring contaminants, such as arsenic and radon, may be present in groundwater at concentrations of potential concern for human health in relatively undeveloped settings that otherwise may not be perceived as susceptible to contamination. Even though the median depth to groundwater in Oahu is more than 300 feet, the common occurrence of anthropogenic compounds in groundwater indicates that Oahu has a high susceptibility to contamination.

Nitrate concentrations in groundwater were above the national background concentrations of 1 milligram per liter (mg/L) in all three study areas. In the Columbia Plateau, nitrate exceeded the human-health benchmark of 10 mg/L in 20 percent of the wells sampled. In the Snake River Plain, nitrate exceeded the human-health benchmark of 10 mg/L in 3 percent of the wells sampled. Nitrate can persist in groundwater for years and even decades in the oxygen-rich groundwater of the Columbia Plateau and the Snake River Plain, so prudent groundwater protection measures are critical to protect drinking water resources by reducing nitrate leaching from the land surface.

Nitrate logistic regression models indicated that areas with a high percentage of land in crops (such as potatoes or sugarcane) and soils with low amounts of organic matter are most likely to have elevated nitrate concentrations in the groundwater. Areas where agricultural activities were absent had much lower probabilities of detecting elevated nitrate concentrations. The Columbia Plateau had a much higher probability of having elevated nitrate concentrations, with most of the land area having greater than a 50 percent probability of elevated nitrate concentrations. Oahu and the Snake River Plain had a much lower probability of having elevated nitrate concentrations because of their lower percentage of agricultural land.

Pesticides were detected at many sites in groundwater of the Columbia Plateau, Snake River Plain, and Oahu but generally at low concentrations below human-health benchmarks. Atrazine and its degradate (a compound produced from the breakdown of a parent pesticide), deethylatrazine, were the most commonly detected pesticides in groundwater sampled in the Columbia Plateau and Snake River Plain. Bromacil was the most commonly detected pesticide on Oahu. The other pesticides most commonly detected in the study areas include simazine, hexazinone, metribuzin, diuron, prometon, metolachlor, *p,p'*-DDE, dieldrin, 2-4-D, and alachlor. DDE (a degradate of DDT) and dieldrin are still being detected in groundwater despite having been banned for more than 30 years. Codetection of multiple pesticides in water from a single well was common. The widespread occurrence of pesticides in groundwater in the study areas indicates that the groundwater is highly susceptible to pesticide contamination.

Some pesticides were detected in groundwater samples from all three study areas, but other pesticides were detected only in samples from Oahu, or only in samples from the Columbia Plateau and Snake River Plain. This is because some pesticides (such as atrazine) are broad-spectrum pesticides that are used on many crops in many different areas of the United States. Other pesticides (such as simazine, metribuzin, and metolachlor) are used on row crops (such as potatoes, barley, and alfalfa) grown in the Columbia Plateau and Snake River Plain, but not on pineapple or sugarcane grown in Oahu.

Atrazine logistic-regression models indicate that areas with a high percentage of land in crops (such as potatoes or sugarcane), a low percentage of fallow land, and highly permeable soils with low amounts of organic matter are most likely to have atrazine detected in the groundwater. Areas where agricultural activities were absent had much lower probabilities of atrazine being detected. The Snake River Plain had a much higher probability of atrazine detections, with more than 50 percent of the land area having greater than a 50 percent probability of atrazine contamination. Oahu had a much lower probability of atrazine contamination, with only 24 percent of the land area having greater than a 50 percent probability of atrazine contamination.

Oahu and the Columbia Plateau had some of the highest percentages of soil fumigant detections in groundwater in the United States. Soil fumigants are volatile organic compounds (VOCs) used as pesticides, which are applied to soils to reduce populations of plant parasitic nematodes (harmful rootworms), weeds, fungal pathogens, and other soil-borne microorganisms. They are used in Oahu and the Columbia Plateau on crops such as pineapple and potatoes. All three areas (Columbia Plateau, Snake River Plain, and Oahu) had fumigant concentrations exceeding human-health benchmarks for drinking water.

Introduction

Since 1991, the National Water-Quality Assessment (NAWQA) program of the U.S. Geological Survey (USGS) has measured water-quality status and trends in major aquifer systems throughout the United States (Gilliom and others, 1995). Pilot efforts were undertaken as early as 1986 in some areas in the United States (Hirsch and others, 1988). Part of the NAWQA sampling effort included sampling of basaltic aquifers in Washington, Idaho, and Hawaii, where these important aquifers serve a population of more than two million people and billions of dollars of agricultural industry. Water-quality data used in this study were collected to assess the effects of primary land use and hydrologic conditions on the concentration and distribution of anthropogenic and naturally occurring compounds in shallow groundwater within individual NAWQA study units (Gilliom and others, 1995).

In most groundwater basins, the quality of shallow groundwater is influenced over relatively short time scales by near-surface activities and, therefore, can be used as an indicator of land-use effects on shallow aquifers (Barbash and Resek, 1996). Although shallow aquifers are not typically used as municipal drinking-water supplies, domestic wells, which are typically untreated, frequently withdraw water from these aquifers for drinking and other household use. National studies have indicated that nonpoint chemical contamination of groundwater in urban and agricultural land-use settings is occurring (U.S. Geological Survey, 2001).

Purpose and Scope

The objective of this report was to evaluate the effect of agricultural and urban land uses on groundwater quality in three areas with similar aquifer properties: Columbia Plateau basaltic-rock aquifers, Snake River Plain basaltic-rock aquifers, and Hawaiian volcanic-rock aquifers (only the island of Oahu was studied in Hawaii so that project resources and funding could be concentrated on the most populous island, where water-quality issues have the greatest effect on society). The groundwater quality of sedimentary aquifers (basin-fill aquifers) that overlie basalts of the Columbia Plateau and Snake River Plain was also investigated. On Oahu, sediments overlie volcanic-rock aquifers along the coastal perimeter of the island, but the sedimentary aquifers are not used as drinking-water sources, so they were not investigated for this report. Groundwater-quality data collected from 1992 to 2010 from domestic, monitoring, and public-supply wells in Washington, Idaho, and Oahu were used. This report summarizes water-quality data for nutrients, pesticides (and their degradation products), and VOCs and compares these data to human-health benchmarks where applicable. Potential explanatory factors influencing shallow groundwater quality, such as aquifer oxidation-reduction (redox) conditions, land use, general soil characteristics, and irrigation practices, are identified.

Environmental and Hydrogeologic Setting in the Columbia Plateau, Snake River Plain, and Hawaiian Principal Aquifers

Understanding the natural environmental and hydrogeologic system, including geochemical processes and effects of human activities on groundwater, is essential for assessing the vulnerability of groundwater resources to contamination. Groundwater quality in the Columbia Plateau, Snake River Plain, and Hawaiian principal aquifers (fig. 1)

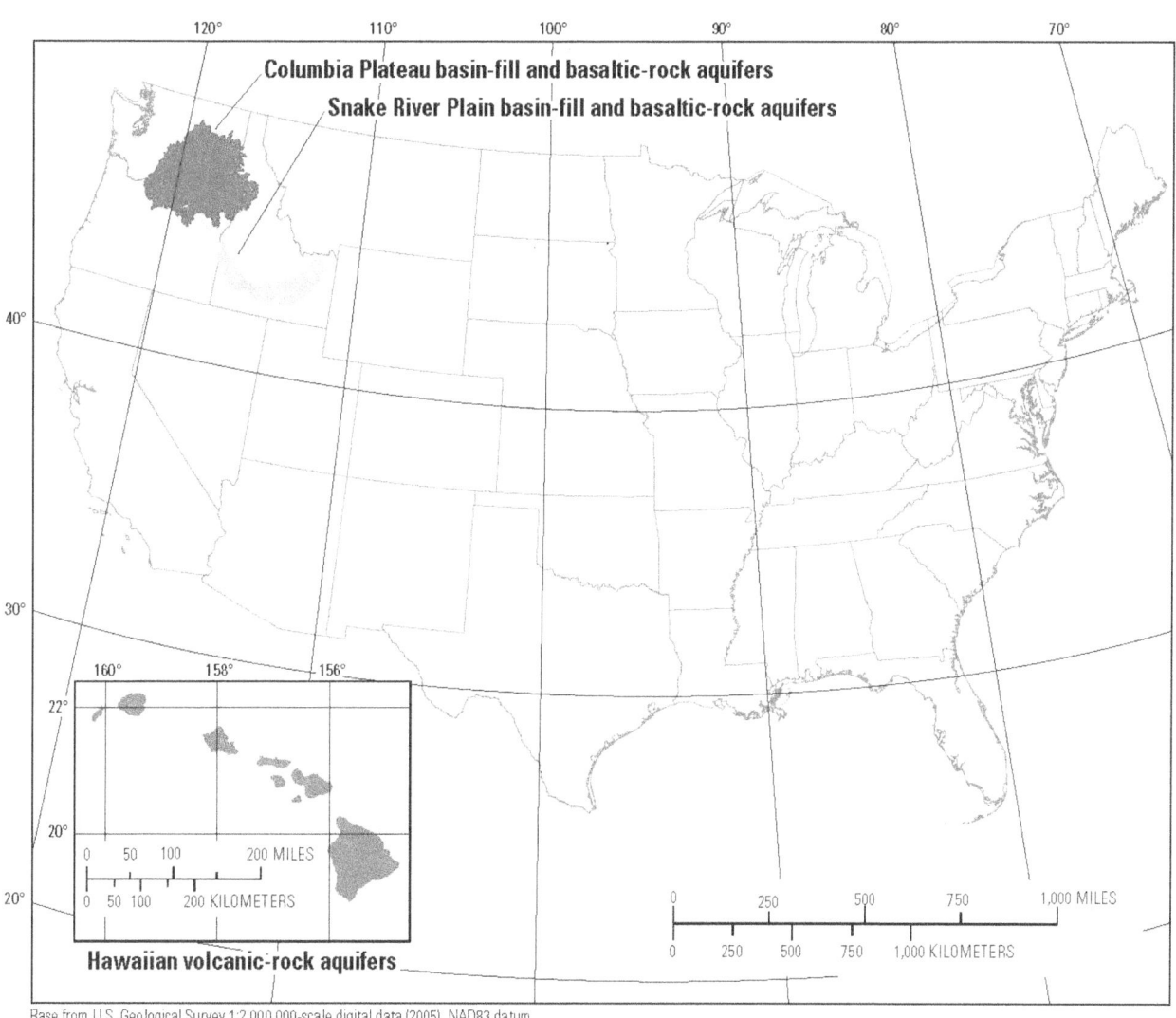

Base from U.S. Geological Survey 1:2,000,000-scale digital data (2005), NAD83 datum.
Albers Equal Area Conic projection, standard parallels 29°30'N and 45°30'N,
central meridian 96oW (continental US), 8°N and 18°N, central meridian 157°W (Hawaii).

Figure 1. Location of principal aquifers in the Columbia Plateau, Snake River Plain, and Hawaii.

is affected by a variety of natural and human influences. Natural features, such as rock type and geochemical conditions, affect the movement of chemical constituents in groundwater and the potential for degradation of constituents, and can render groundwater vulnerable to contaminants introduced at the land surface in agricultural and urban settings. Natural hydrologic cycles have been modified heavily by river diversion, pumping, and irrigation—in volumes that are comparable to natural hydrologic flows.

This section summarizes background information for the Columbia Plateau, Snake River Plain, and Hawaiian principal aquifers and provides the context for understanding findings about water quality in these hydrologic systems. Topics include the environmental setting, population, land use, water use, and hydrogeologic setting.

Environmental Setting

The Columbia Plateau, Snake River Plain, and Hawaiian Islands are large volcanic provinces in the western United States and mid-Pacific Ocean (fig. 2). The three areas have been grouped here for study as the "Western Volcanics principal aquifers" because each contains extensive regional aquifers of a hard, gray, volcanic rock called basalt. Columbia Plateau aquifers encompass 58,000 square miles in parts of Washington, Oregon, and Idaho (Miller, 1999). Snake River Plain aquifers encompass 14,400 square miles across southern Idaho and a small sliver of eastern Oregon. The main Hawaiian Islands encompass 6,440 square miles of land area at the southeast end of the Hawaii-Emperor mountain chain, a linear chain of islands and underwater seamounts in the mid-North Pacific Ocean.

The largest and most productive aquifers in these areas are bedrock aquifers of basalt; the chemical makeup of some Hawaiian rocks differs slightly from that of basalt (Oki and Brasher, 2003), so the more general term "volcanic rock" is used there. However, each area also has subordinate sedimentary aquifers and confining units. Sedimentary "basin-fill" aquifers have been mapped in the Columbia Plateau and Snake River Plain, where basalts and sediments accumulated in subsiding basins to form complex, multilayered aquifer systems. Hawaii also has sedimentary deposits but they are of lesser importance for water supply, and nearly all groundwater there is developed from the more productive volcanic-rock aquifers.

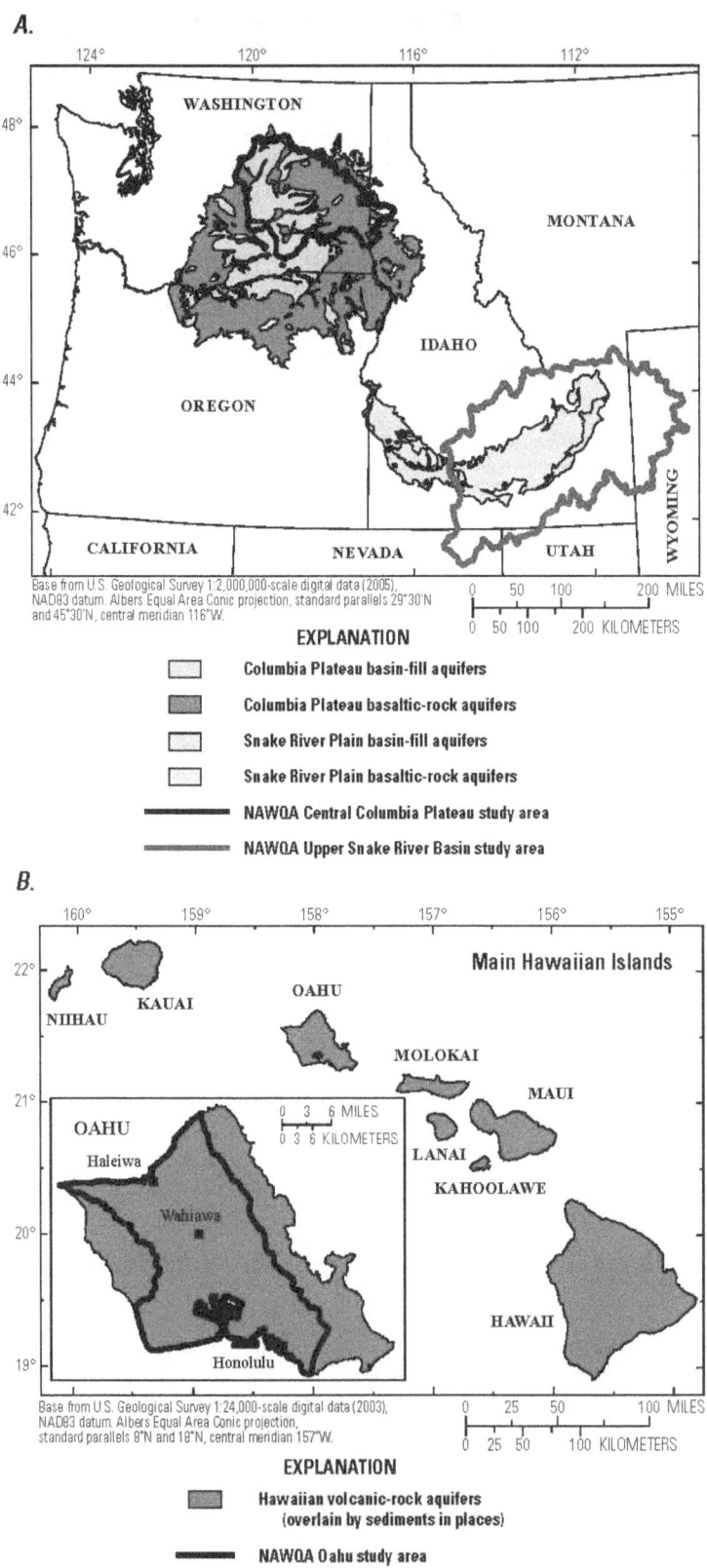

Figure 2. Principal aquifers of (A) the Columbia Plateau and Snake River Plain and (B) the eight main Hawaiian islands at the southeast end of a mid-Pacific mountain chain. Inset shows Oahu in greater detail.

Climate in the Columbia Plateau and Snake River Plain is temperate and arid to semiarid, with mean annual precipitation less than 15 inches over large areas and less than 10 inches in places. Precipitation is greater in adjacent mountains and tributary valleys, as much as 45–60 inches or more (Nelson, 1991; Clark and others, 1998). Climate in Hawaii is subtropical, and precipitation varies strongly with altitude and exposure to prevailing easterly trade winds, from semiarid in leeward lowlands to humid at higher windward elevations where annual rainfall exceeds 200 inches on several islands (Giambelluca and others, 1986).

Population

Oahu is the most populous and urbanized of the three areas (fig. 3), with a population of 902,000 in 2005, most of it concentrated in urban Honolulu and nearby suburban communities in southern Oahu (U.S. Census Bureau, 2008). Population in the Columbia Plateau was nearly as great, at 821,000 in 2005, spread among farming communities across the Plateau but also concentrated in urban centers such as the tri-city area that includes Pasco and its adjacent sister cities of Richland and Kennewick. Population in the Snake River Plain was 491,000 in 2005 and, as in the Columbia Plateau, was spread among rural farming communities and urban centers such as Boise, Twin Falls, Pocatello, and Idaho Falls. Population has increased steadily in all three areas, more than

doubling in the Snake River plain since 1970 (factor of 2.2) and increasing by factors of 1.6 in the Columbia Plateau and 1.4 on Oahu during the same time interval.

Land Use

Land use and land cover are diverse in the Western Volcanics region, and groundwater is vulnerable to effects of overlying urban and suburban development, as well as agricultural management practices such as fertilizer and pesticide application at the land surface. Main types of land use and land cover include forest and rangeland, extensive agricultural areas, and urban centers of various extent and density (Homer and others, 2004) (fig. 4). Principal crops in the Columbia Plateau and Snake River Plain include alfalfa, cereal grains (wheat, barley, oats), potatoes, field crops, and corn; the Columbia Plateau also has orchards (primarily apples and cherries) and vineyards. Hawaii was dominated historically by two crops, sugarcane and pineapple. Sugarcane was phased out in 1996 on Oahu, leaving pineapple and various field crops. Agricultural land use has remained prevalent in the Columbia Plateau and Snake River Plain, but has diminished in Hawaii in recent decades. Former agricultural lands in central and southwest Oahu have been converted to suburban and light commercial use since the 1960s.

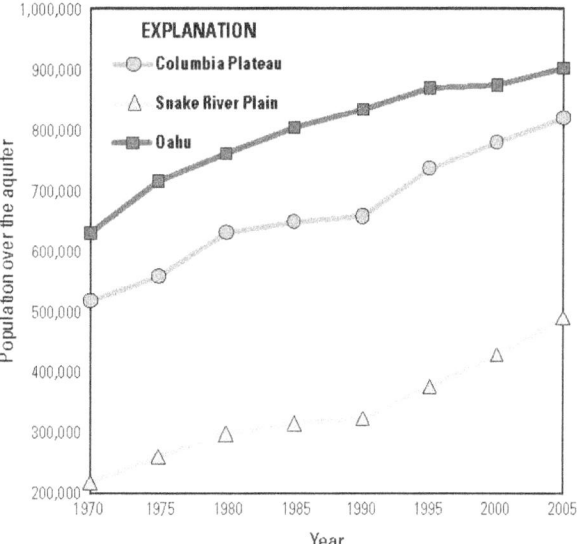

Figure 3. Population of the Columbia Plateau, Snake River Plain, and Oahu study areas (modified from U.S. Census Bureau, 2008).

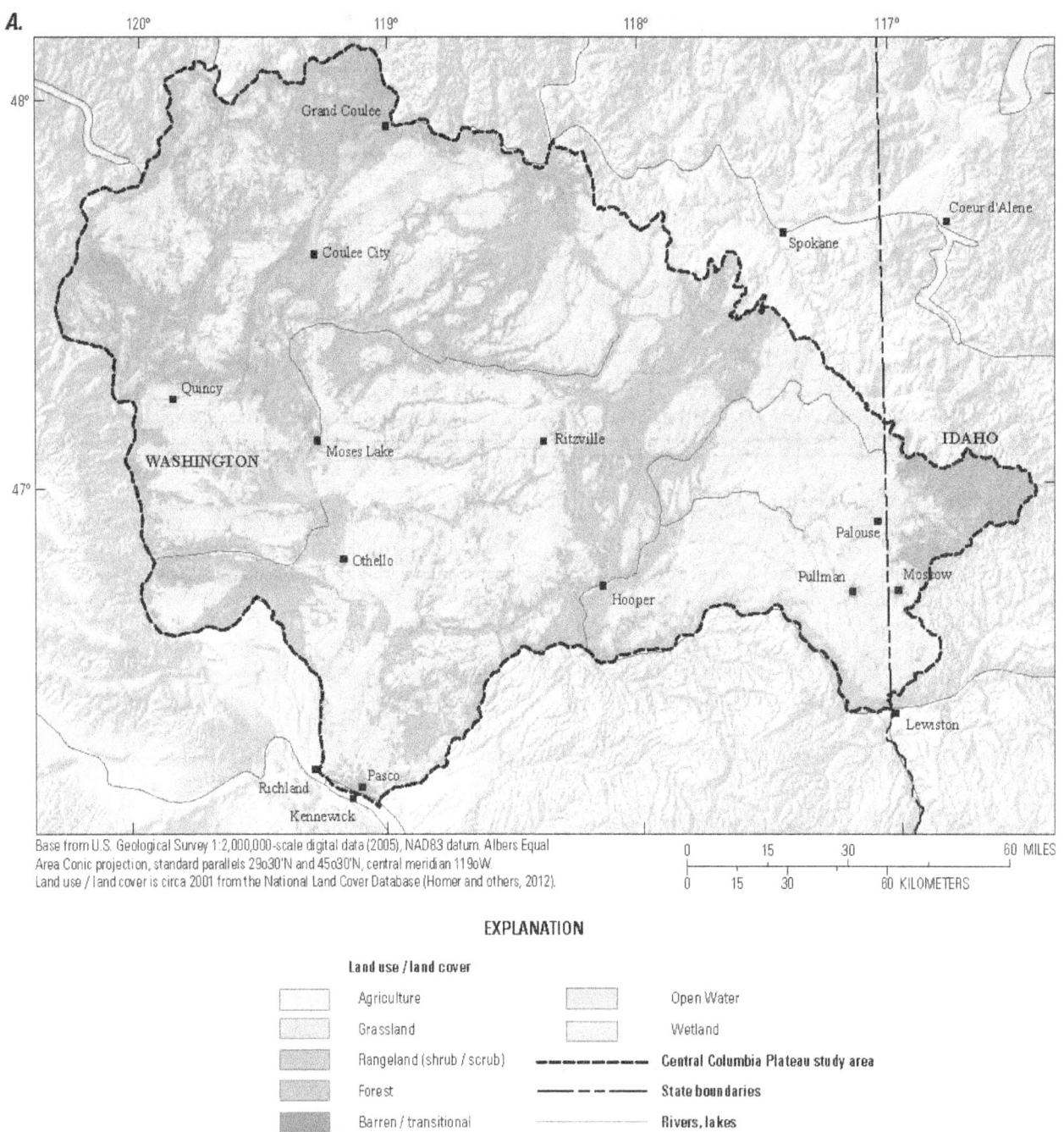

Figure 4.   Land use for 2001 in (*A*) the Columbia Plateau, (*B*) the Snake River Plain, and (*C*) Hawaii.

B.

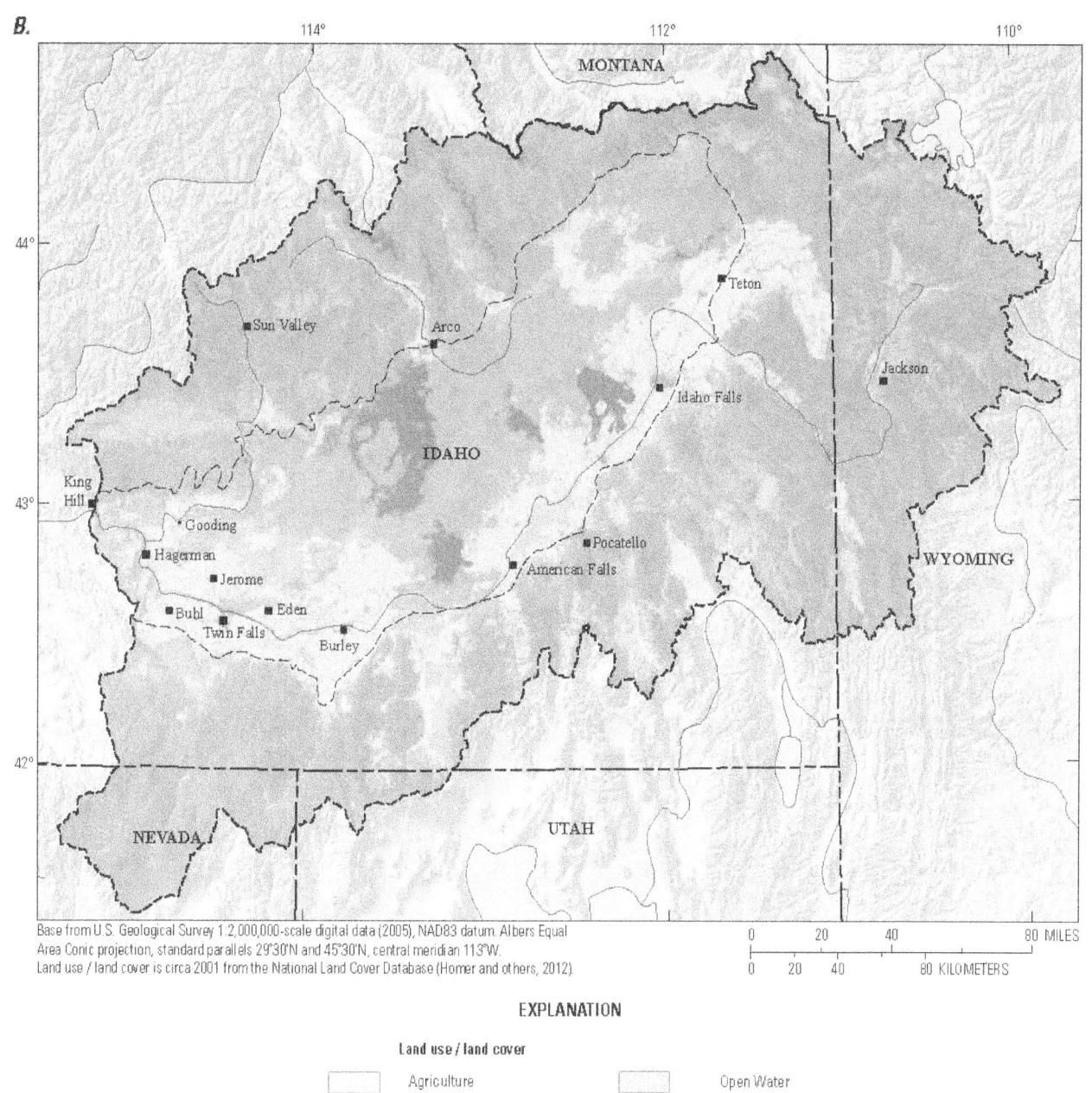

Base from U.S. Geological Survey 1:2,000,000-scale digital data (2005), NAD83 datum. Albers Equal Area Conic projection, standard parallels 29°30'N and 45°30'N, central meridian 113°W. Land use / land cover is circa 2001 from the National Land Cover Database (Homer and others, 2012).

EXPLANATION

Land use / land cover

	Agriculture		Open Water
	Grassland		Wetland
	Rangeland (shrub / scrub)	— — — — —	Eastern Snake River Plain
	Forest	— — — — —	Upper Snake River Basin study area
	Barren / transitional	— — — — —	State boundaries
	Urban		Rivers, lakes

Figure 4.—Continued.

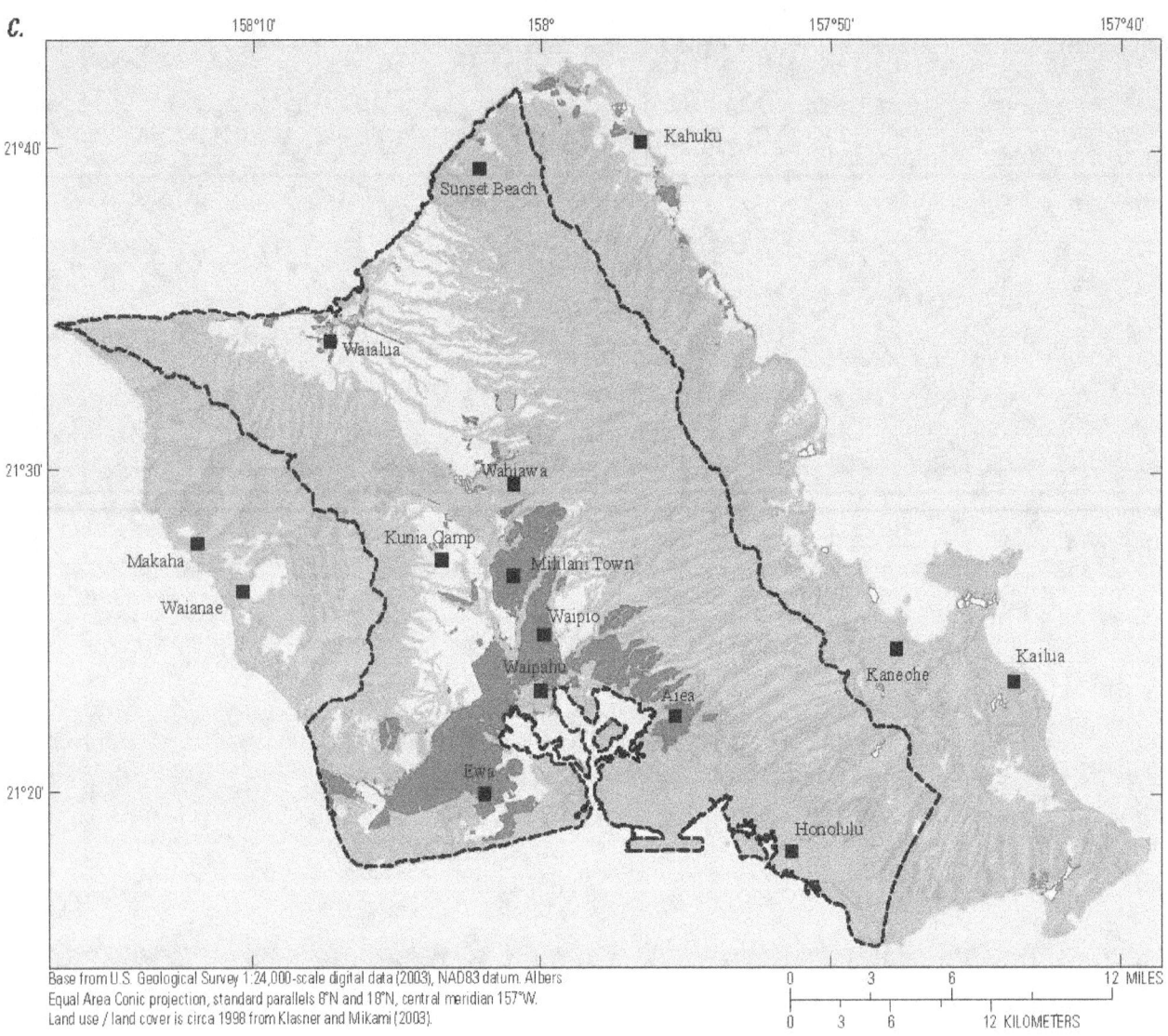

Base from U.S. Geological Survey 1:24,000-scale digital data (2003), NAD83 datum. Albers
Equal Area Conic projection, standard parallels 8°N and 18°N, central meridian 157°W.
Land use / land cover is circa 1998 from Klasner and Mikami (2003).

EXPLANATION

Land use / land cover

- Agriculture
- Urban
- Urban, converted from agricultural use from 1950 to 1998
- Other (mostly forest)
- ------- NAWQA Oahu study area

Figure 4.—Continued.

Water Use

Water in the Western Volcanics is supplied principally by groundwater and by surface-water diversion projects on the Columbia and Snake Rivers and on smaller streams in Hawaii. Crop irrigation is the predominant use of groundwater in the Snake River Plain and Columbia Plateau (Maupin and Barber, 2005) (fig. 5), whereas public supply is the largest use of groundwater from Hawaiian volcanic-rock aquifers. Snake River Plain basaltic-rock aquifers supply the largest amount of groundwater by far: 2,500 million gallons per day for irrigation alone. Although domestic wells provide drinking water for a significant portion of the population in the Columbia Plateau and Snake River Plain, the actual amount of water pumped by a domestic well is small, so the overall water use is small.

The ways in which water is used can affect water quality. Diversion of river water to irrigated fields can concentrate dissolved minerals and salts through evaporation. The same mechanism acts on groundwater pumped to the surface for irrigation. Reinfiltration of that water to be withdrawn again by wells can set up a cycle of recirculation in which dissolved constituents such as salts and nitrate become progressively more and more concentrated through multiple evaporation cycles (Rupert, 1997). Urban use of water also concentrates dissolved constituents in comparison to initial source waters. Although municipal wastewaters are treated extensively to remove biological wastes and nutrients such as nitrate, discharged final effluent is higher in dissolved constituents than source waters and contains urban chemicals, such as detergents, pharmaceuticals, and other household products and their degradates. Municipal effluents are generally discharged to streams and rivers and also to ocean outfalls in Hawaii.

Hydrogeologic Setting

The largest and most productive aquifers in the Columbia Plateau, Snake River Plain, and Oahu are composed of basalt, but each area also has smaller sedimentary aquifers. Sedimentary "basin-fill" aquifers occur in the Columbia Plateau and Snake River Plain, where basalts and sediments accumulated in low-lying basins to form complex, multilayered aquifer systems. Hawaii also has sedimentary deposits, but they are a less important water supply, and nearly all groundwater is pumped from the more productive basalt aquifers.

The basalt and sedimentary aquifers are highly vulnerable to contamination by chemicals applied, spilled, or disposed at the land surface because of the physical characteristics of the basalt flows. Individual basalt flows have three zones: (1) a top layer of rock fragments (known as rubble or "aa clinker" in Hawaii, and as "flow-top breccia" in Washington and Idaho), (2) a hard, dense, massive central layer, and (3) a basal layer

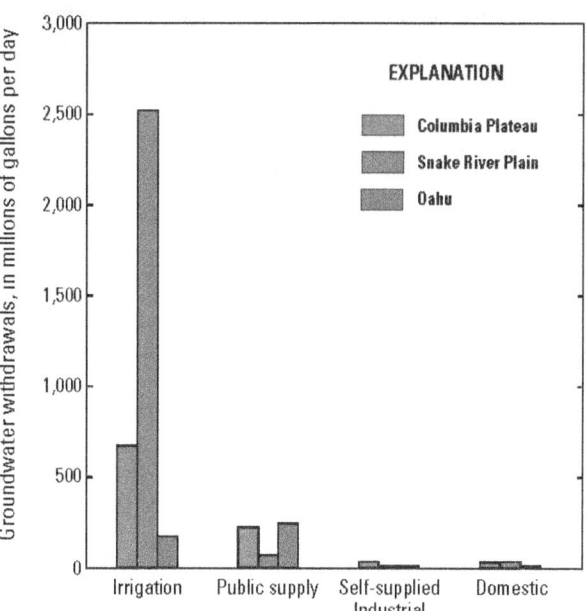

Figure 5. Water use in 2000 for the Columbia Plateau and Snake River Plain basaltic-rock aquifers and Hawaiian volcanic-rock aquifers.

of rock fragments. The top and bottom layers are composed of rubble because those parts of the lava flows cool very quickly and become fragmented; the middle layer is very hard and dense because it cools much more slowly. Basalt aquifers are highly vulnerable to contamination because the rubble zones allow large amounts of groundwater and contaminants to flow unimpeded, and the chemical composition of basalt is inert to most contaminants.

Although the Western Volcanics aquifers share a common rock type (basalt), the thickness and geometry of the basalt layers differ in important ways as a result of their depositional environments and eruptive habits. Columbia River and Snake River basalts erupted within subsiding basins that contained major rivers and lakes. As a result, these basins contain sequences of basaltic-rock aquifers interlayered with sedimentary basin-fill aquifers and confining units. Because lava flows spread out over relatively flat terrain within the basins, lava ponded to form fairly flat-lying and thick flows, particularly in the Columbia Plateau, where extremely high-volume fissure eruptions produced individual lava sheets 30 to 300 feet thick extending over several thousand square miles (Tolan and others, 2009). Basaltic lavas in the Snake River Plain emanated from fissures and gently domed shield volcanoes at lower volumetric rates, typically extending over 50 to 100 square miles and ranging in thickness from a few feet to more than 100 feet (Whitehead, 1992), averaging on the order of 20 to 25 feet (Mundorff and others, 1964).

Hawaiian lava flows are the thinnest among the three areas, most commonly several feet to 30 feet thick. Most lavas flowed down the flanks of domed shield volcanoes, typically at slopes of 3 to 20 degrees. As a result, most Hawaiian flank flows are long, narrow tongues that may be miles long but tend to be less than a mile wide over much of their length. Lavas did spread out areally and pond to greater thickness in some places (where terrain was flat or where lava filled erosional basins or calderas), but the most productive Hawaiian aquifers consist of thin-bedded flank flows. Sedimentary aquifers and confining units are present, but volcanic rocks form the main drinking-water aquifers; where sediments play an important role it is mostly as confining units to the volcanic-rock aquifers.

Thin-bedded basalts, such as the Hawaiian lava flows, contain open voids, tubes, and rubble (flow-top breccia) that impart high permeability and ease of lateral groundwater flow. Layers of dense rock contain fractures that allow cross-layer flow, but at lesser rates than flow through the more permeable breccia and voids. Thicker basalts, such as the Columbia Plateau basalts, tend to have thicker dense layers and less breccia. A single Columbia River basalt flow 50–100 feet thick might be 90 percent dense rock and only 10 percent breccia. Vertical fractures in the dense layer allow some water movement, but not nearly as readily as the breccia layers between flows. These "interflow" zones of breccia (and permeable sediment deposited between some flows) allow the greatest water movement and yield the largest amounts of water to wells that penetrate them, whereas the dense rock layers tend to act as confining units in the Columbia

Plateau. Interflow sediments can also be confining units if the sediments are fine grained. Snake River basalts are intermediate between thick-bedded Columbia River basalts and thin-bedded Hawaiian basalts, with permeability and ease of groundwater flow closer to Hawaiian basalts than to Columbia River basalts.

Aquifer Systems and Groundwater Flow Systems

The Columbia Plateau is considered a "multiaquifer system"—a layered sequence of aquifers that allow flow from aquifer to aquifer, but with some resistance from weak confining units of fine-grained sediments and weathered rock between the aquifers (fig. 6). In the Columbia Plateau aquifer system, groundwater flow is predominantly lateral over great distances, as much as a hundred miles (Lindholm and Vaccaro, 1988). However, water also infiltrates to deeper aquifers in recharge areas (right side of fig. 6) and upward to shallower aquifers where the regional flow discharges, such as to the Columbia River (left side of fig. 6). Columbia River basalts are estimated to reach a maximum thickness of as much as 15,000 feet (Miller, 1999). Several major basalts have been mapped as stratigraphic formations and are considered to constitute individual aquifers (fig. 7), distinguishable by water-level differences in wells that tap the aquifers (Kahle and others, 2011). For example, groundwater flow in the Grande Ronde Basalt of the Columbia Plateau aquifer system is generally east to west, and varying in direction where flow converges to discharge to the Columbia and Snake Rivers.

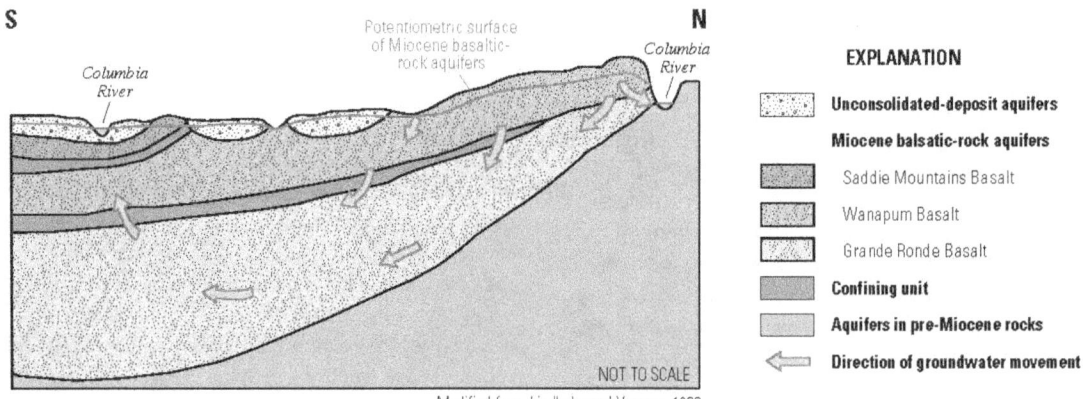

Modified from Lindholm and Vaccaro, 1988

Figure 6.   Generic geologic cross-section of the Columbia Plateau. Modified from Lindholm and Vaccaro (1988).

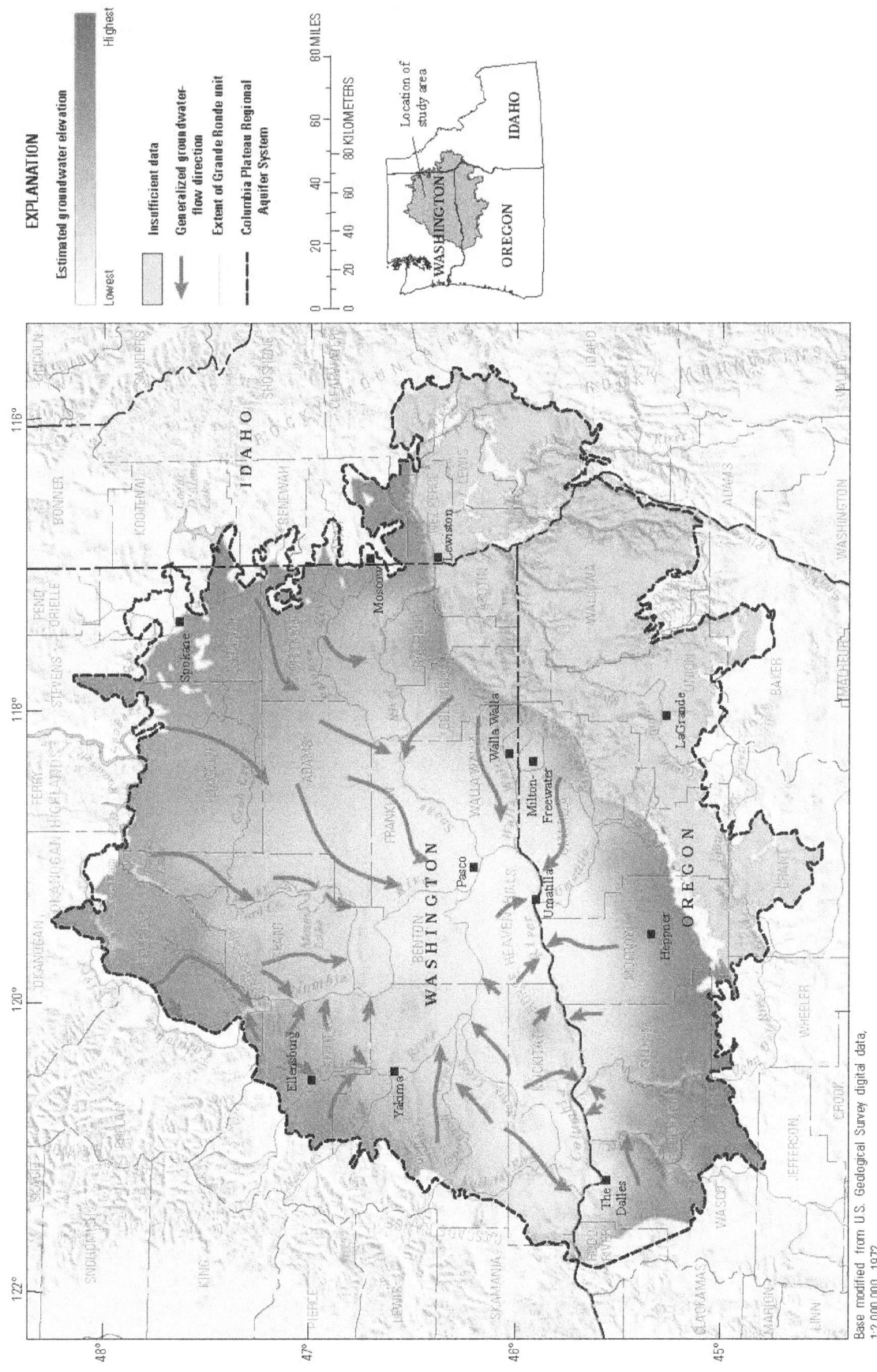

Base modified from U.S. Geological Survey digital data,
1:2,000,000, 1972

Figure 7. Generalized groundwater levels and directions of lateral groundwater movement for the Grande Ronde unit, Columbia Plateau Regional Aquifer System, Washington, Oregon, and Idaho (modified from *Vaccaro*, 1999 and Snyder and others, 2010).

In the Eastern Snake River Plain and Hawaiian aquifers, regionally extensive confining units are not recognized within the volcanic-rock aquifers as much as in the Columbia Plateau. In many places only a single basaltic-rock aquifer is recognized within depths penetrated by wells for practical purposes such as groundwater withdrawal. The most productive aquifers consist of thick accumulations of thin-bedded lavas that are hydraulically well connected over large distances, with little to impede downward infiltration to the deep water table in most places, even though the water table may lie several hundred feet below land surface. Basalts are estimated to reach a maximum thickness of 5,500 feet in the Eastern Snake River Plain (Miller, 1999), where groundwater flows to the southwest and discharges west of Twin Falls where the basaltic-rock aquifer pinches out (fig. 8). Groundwater flow in the Snake River Plain aquifer system is perpendicular to potentiometric (water level) contours, mostly northeast to southwest in the Eastern Plain and in various directions in the Western Plain. Much of the discharge from the aquifer system is to the Snake River. Potentiometric contours are widely spaced where the Eastern Snake River Plain aquifer is thickest and most permeable. Areas of perched water are underlain by low-permeability material, typically lakebed sediments.

In the Eastern Snake River Plain, multi-aquifer relationships between basalts and sediments are recognized locally near the margins of the plain, where streams washed sediments onto the lava plain throughout its eruptive history (fig. 9). In the Western Plain, basalts are more discontinuous and are interbedded with sediments throughout the plain to form a complex multi-aquifer system. Perched water accumulates where sediments are fine grained and low in permeability, the best example being lakebed sediments deposited when lava flows dammed the Snake River to form lakes at various times.

Hawaiian volcanic-rock aquifers extend as deep as the shield volcanoes (to the adjacent ocean floor about 16,000 feet below sea level), but freshwater is withdrawn only down to a maximum depth of about a thousand feet below sea level. The central Oahu groundwater flow system is the largest water resource on Oahu and is the area studied by NAWQA (fig. 10). Groundwater converges into a central plateau between the east and west mountain ranges and then diverges north and south.

Fresh groundwater on Oahu extended as much as 1,600 feet below sea level, as lenses above denser salt water, before groundwater withdrawal lowered water tables and depleted the freshwater lenses within the aquifer. Groundwater is impounded to heights of several hundred feet above sea level by intrusive volcanic dikes along the mountain axes, but the larger and more productive water resources are the thick freshwater lenses within dike-free flank lavas (fig. 11). Coastal sediments confine the volcanic rock aquifer and build up the freshwater lens to greater thickness and water-table height than where sediments are absent. Small bodies of perched water occur on ash beds and above dense valley-filling basalts.

Multi-aquifer relationships have been recognized between Hawaiian shield volcanoes, for example between the Waianae and Koolau Volcanoes on Oahu (Hunt, 1996) and between Mauna Kea and Mauna Loa Volcanoes on the island of Hawaii (Thomas and others, 1996). Multi-aquifer relationships and distinct water-level differences with depth also are commonly recognized locally where younger lavas have filled previously eroded terrain and within stratified sequences of coastal sedimentary aquifers and confining units (Hunt, 1996).

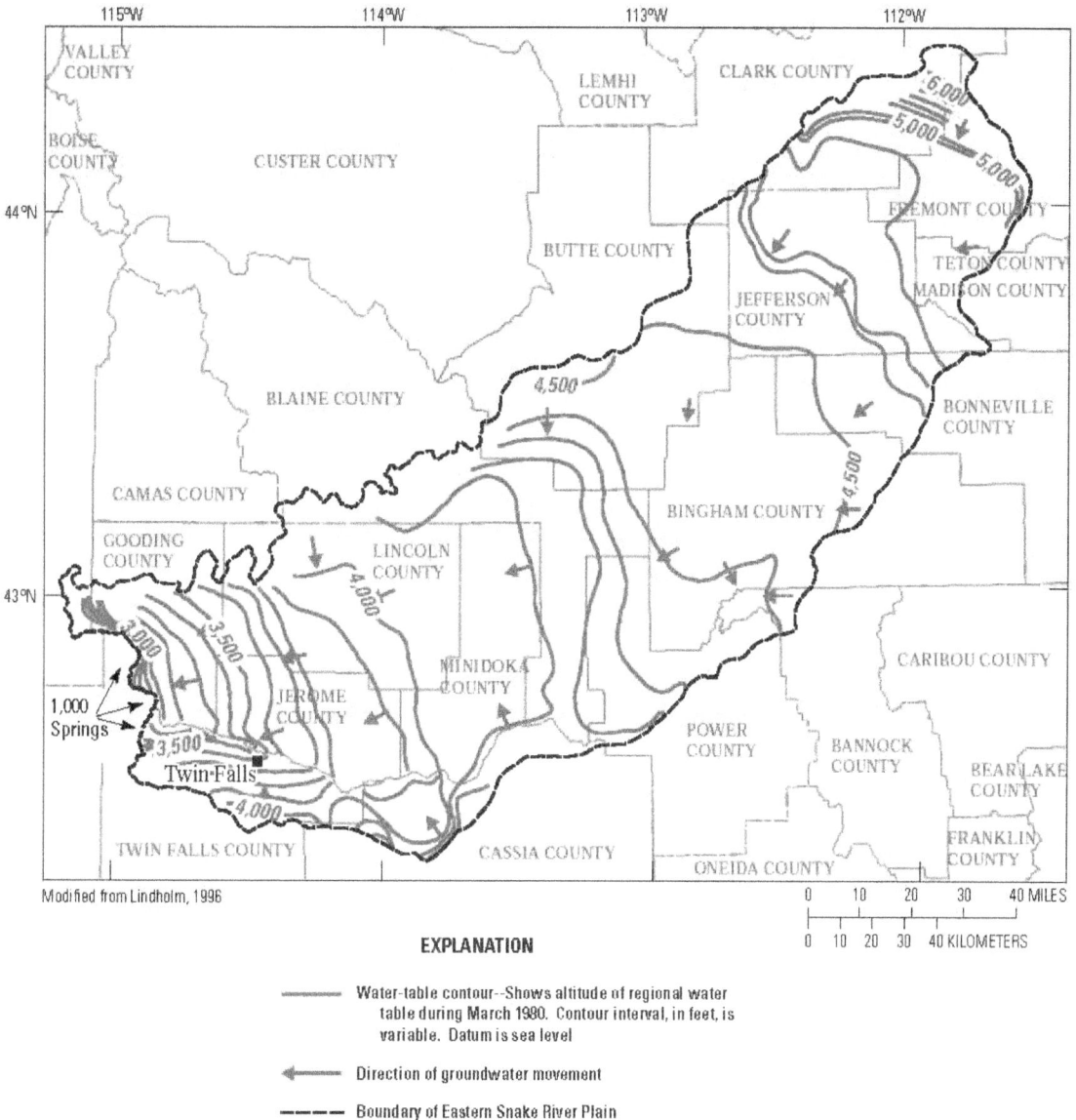

Figure 8. Groundwater flow in the Snake River Plain aquifer.

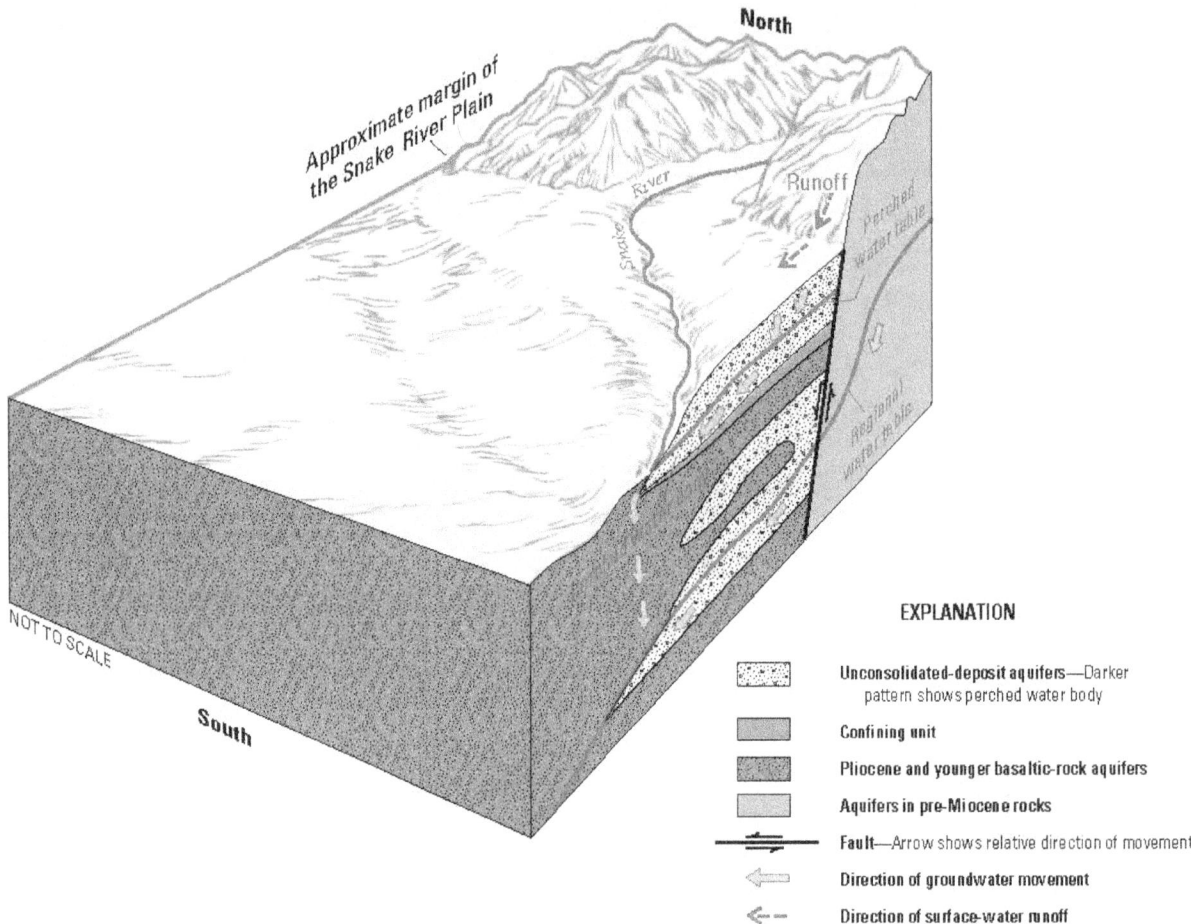

Figure 9. Block diagram and generic geologic cross-section of the Snake River Plain. Modified from Miller (1999).

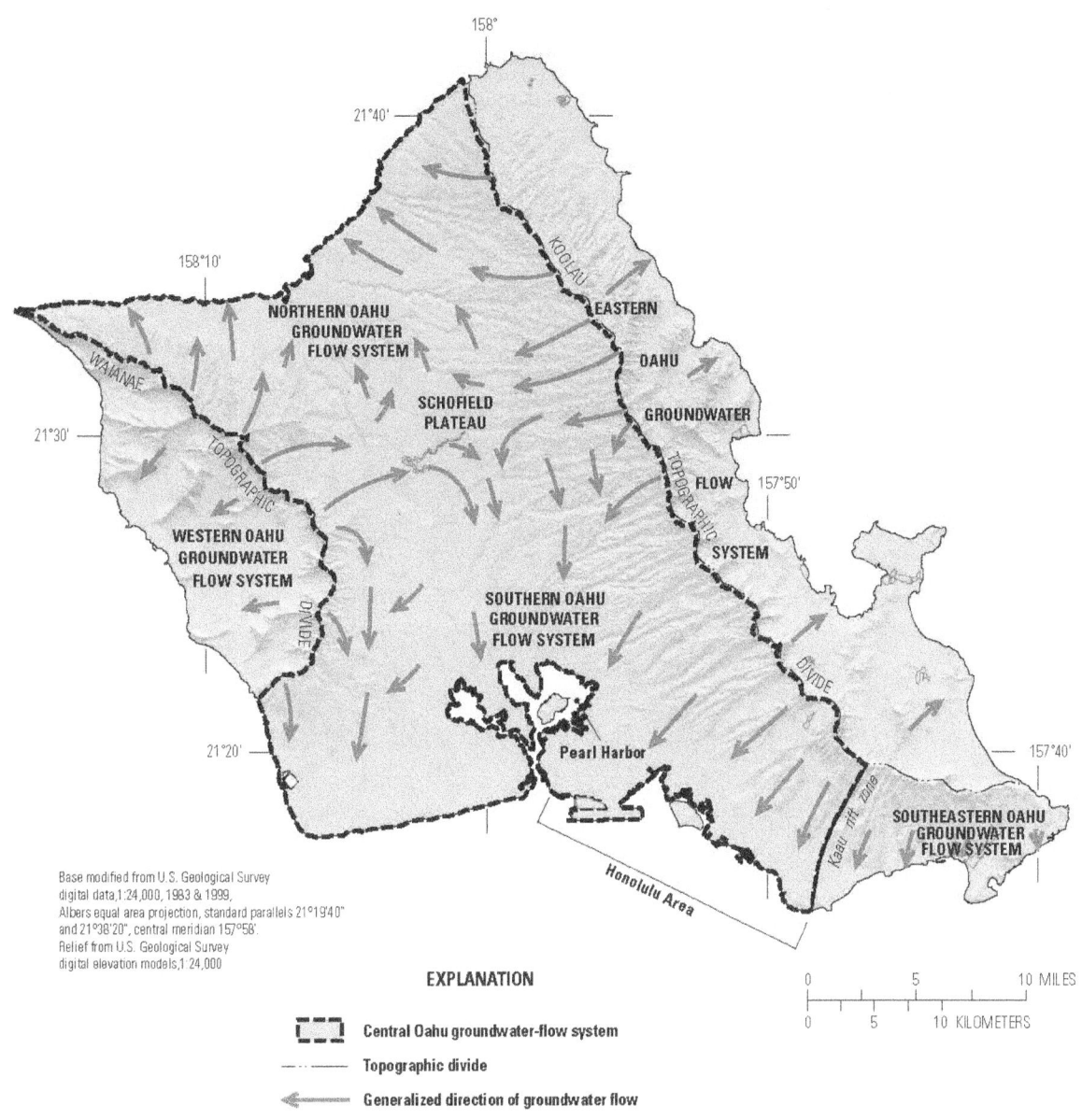

Base modified from U.S. Geological Survey
digital data,1:24,000, 1983 & 1999,
Albers equal area projection, standard parallels 21°19'40"
and 21°38'20", central meridian 157°58'.
Relief from U.S. Geological Survey
digital elevation models,1:24,000

EXPLANATION

⬚ Central Oahu groundwater-flow system

---·--- Topographic divide

⬅ Generalized direction of groundwater flow

Figure 10. Groundwater-flow directions on Oahu, Hawaii.

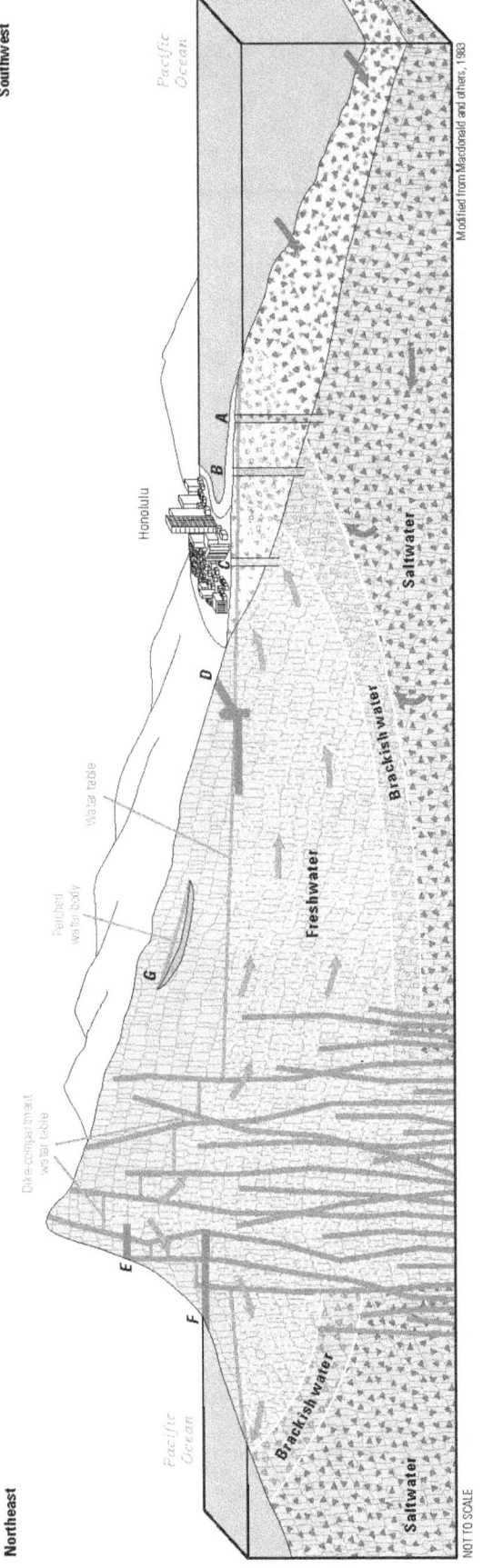

Modified from Macdonald and others, 1983

EXPLANATION

	Sedimentary deposits (caprock)
	Volcanic rocks
	Volcanic ash
	Dike
E	Tunnel or shaft
B	Well

Direction of groundwater movement

Freshwater

Saltwater

Fresh groundwater moves from topographically high areas toward the ocean. Fresh groundwater flow is generally downward in inland areas, upward in the coastal areas, and horizontal in between. A saltwater circulation exists beneath the freshwater lens: saltwater flows landward in deeper parts of the aquifer, rises, then mixes with fresher water and discharges to the ocean.

A freshwater-lens system underlies much of southern Oahu. Well *A* produces saltwater from below the transition zone, well *B* produces brackish water from the transition zone, and well *C* produces freshwater. Horizontal tunnel *D* (sometimes called a Maui shaft) produces large volumes of freshwater by skimming water from near the top of the freshwater lens. Tunnel *E* (sometimes called a Lanai shaft) is dug horizontally into one or more of the dike-bounded compartments. Location *F* indicates a perched water body containing minor amounts of water. A bed of almost impermeable ash creates a perched water body at site G.

Figure 11.   Hypothetical saltwater-freshwater interaction on Oahu, Hawaii.

Methods

General Study Design for Assessing Water Quality in the Columbia Plateau, Snake River Plain, and Oahu Principal Aquifers

This NAWQA study provides a systematic assessment and analysis of water quality in the principal aquifers of the Columbia Plateau, Snake River Plain, and Oahu. An approach was developed that assessed overall water quality in the aquifer systems and also enabled an understanding of the linkages between the quality of water recharging the aquifer system, effects on water quality during transport through the hydrologic system, and the quality of the water resource used for human consumption or discharged to surface-water bodies and used for recreation or supporting ecological communities and economies.

NAWQA groundwater studies were conducted in selected areas of the principal aquifers (fig. 12). The Central Columbia Plateau studied the northern portion of the larger Columbia Plateau system. In the Snake River Plain, only the Eastern Plain was studied (along with its tributary drainage areas) because of its regionally continuous basaltic-rock aquifer (basalts in the Western Plain are discontinuous and interbedded with sediments in a less productive aquifer system). In Hawaii, only the island of Oahu was studied by NAWQA.

Study Components of the Columbia Plateau, Snake River Plain, and Oahu Principal Aquifers

The design of groundwater studies in the Central Columbia Plateau, Snake River Plain, and Oahu considered the distribution of forested, agricultural, urban, and suburban land use on the landscape. Undeveloped areas, such as forested lands, have few sources of contaminants and thus were not sampled in proportion to the area that they cover. The major study types included agricultural land-use studies, major-aquifer studies, and special studies focused specifically on parts of the aquifers where particular contaminants were expected to be higher than in other areas. Well networks and their abbreviations are listed in table 1 and are portrayed on maps in figure 12, which was compiled from Clark and others (1998), Williamson and others (1998), and Anthony and others (2004).

Land-Use Studies

Land-use studies are designed to assess the concentrations and distribution of water-quality constituents in groundwater near the water table associated with the most important land-use and hydrogeologic settings in each study area. For the Western Volcanics principal aquifers, the only land-use type targeted was agricultural, although NAWQA groundwater studies in other parts of the Nation also investigated urban land use. For the land-use studies, a network of 20 to 30 monitoring wells that penetrated shallow groundwater in the uppermost part of the aquifer system was installed. In some places (such as the Snake River Plain), where large depths to water or geologic conditions made installation of monitoring wells too costly, existing wells were used. In many instances, the networks targeted a single crop type, irrigation regime, or other key characteristic.

Agricultural land-use studies were conducted in the Columbia Plateau (3 studies; ccptlusag1, ccptlusag2, and ccptlusor1) and Snake River Plain (4 studies; usnkluscr1, usnkluscr2, usnkluscr3, usnkluscr4) (table 1). No land-use studies could be designed for Oahu, where the expense of drilling forced a reliance on existing wells and a close link could not be assured between water samples and overlying land use (depth to water was large and many candidate wells had long open intervals that would not sample just the shallowest, recently recharged water).

Columbia Plateau land-use studies each employed two subnetworks of wells: (A) existing domestic wells tapping sediments and basalt and (B) newly drilled shallow monitoring wells emplaced only in the uppermost unconsolidated sediments. The three Columbia Plateau land-use studies targeted different crops and irrigation regimes: dryland grains, irrigated row crops, and irrigated orchards. Network median depths to water ranged from 16 to 60 feet for the wells used in the land-use studies.

The four Snake River Plain land-use studies all targeted irrigated row crops, but in four different areas along the groundwater flow direction that varied in irrigation type (groundwater or surface water) and depth to water. Snake River Plain land-use studies relied on existing wells because it was too expensive to drill new wells in hard basalt (typical depths to water were 100 feet or more). Sampled wells were mostly domestic, with a few irrigation and stock wells. Network median depths to water were 10 feet for a shallow network in sediments and 150 to 290 feet among the other three networks of wells tapping basalt.

A.

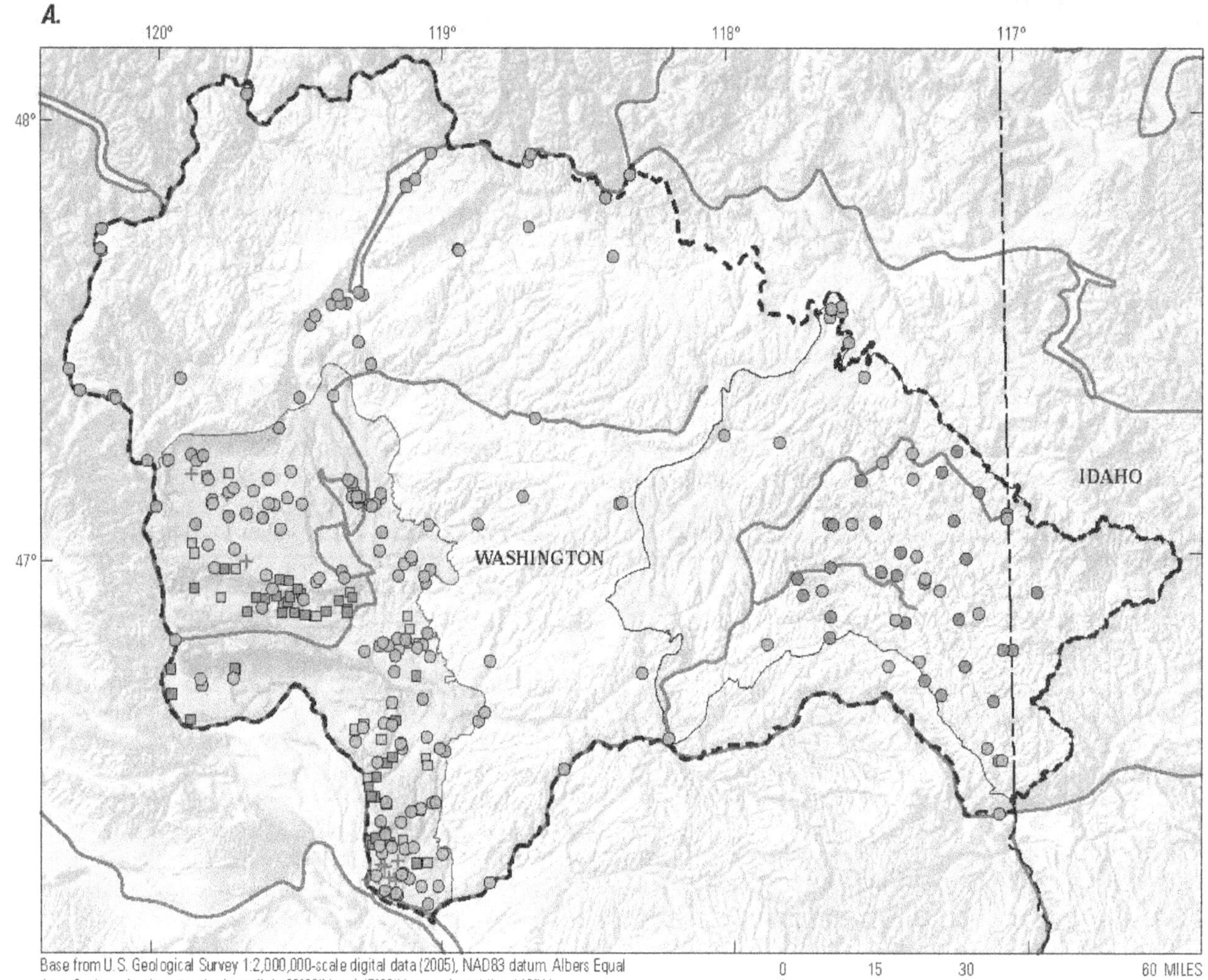

Base from U.S. Geological Survey 1:2,000,000-scale digital data (2005), NAD83 datum. Albers Equal Area Conic projection, standard parallels 29°30'N and 45°30'N, central meridian 119°W.

0 15 30 60 MILES
0 15 30 60 KILOMETERS

EXPLANATION

- Agricultural land-use study, Palouse area - dryland grains (lusag1)
- Agricultural land-use studies, Quincy-Pasco area - irrigated row crops (lusag2) and orchards (lusor1)
- NAWQA Central Columbia Plateau study area. Major-aquifer study (sus1); the entire Central Columbia Plateau study area, including the land-use studies
- State boundaries
- Rivers, lakes
- Wells tapping major aquifers, mixed land uses, public supply and domestic (sus1)
- Wells monitoring groundwater in agricultural areas (dryland farming), domestic wells (lusag1a) and monitor wells (lusag1b)
- Wells monitoring groundwater in agricultural areas (irrigated row crops), domestic wells (lusag2a) and monitoring wells (lusag2b)
- Wells monitoring groundwater in agricultural areas (orchards and vineyards), domestic wells (lusor1a) and monitoring wells (lusor1b)
- Pesticide synoptic network, irrigation drains and sub-surface tile-drain outlets (spcb1)
- EDB synoptic network, shallow domestic wells and irrigation drains (spcb2)

Figure 12. Location of aquifer studies and wells sampled in the (*A*) Columbia Plateau, (*B*) Snake River Plain area, and (*C*) Oahu, 1993–2005.

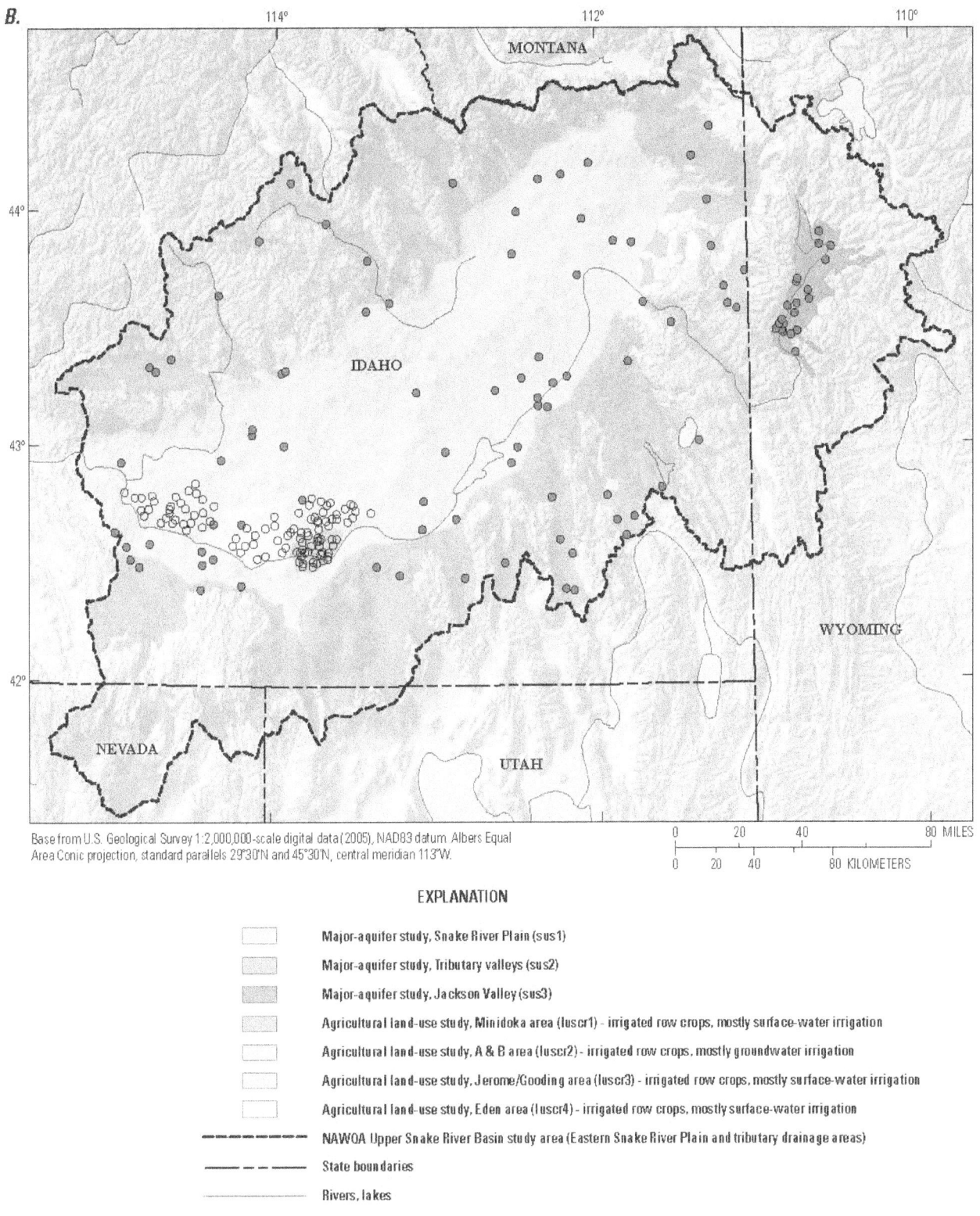

0   20   40   80 MILES

0   20   40   80 KILOMETERS

EXPLANATION

Major-aquifer study, Snake River Plain (sus1)

Major-aquifer study, Tributary valleys (sus2)

Major-aquifer study, Jackson Valley (sus3)

Agricultural land-use study, Minidoka area (luscr1) - irrigated row crops, mostly surface-water irrigation

Agricultural land-use study, A & B area (luscr2) - irrigated row crops, mostly groundwater irrigation

Agricultural land-use study, Jerome/Gooding area (luscr3) - irrigated row crops, mostly surface-water irrigation

Agricultural land-use study, Eden area (luscr4) - irrigated row crops, mostly surface-water irrigation

NAWQA Upper Snake River Basin study area (Eastern Snake River Plain and tributary drainage areas)

State boundaries

Rivers, lakes

Wells tapping major aquifers, mixed land uses - domestic, irrigation, stock, and public-supply wells

Wells in irrigated row crop areas, mostly domestic

Figure 12.—Continued.

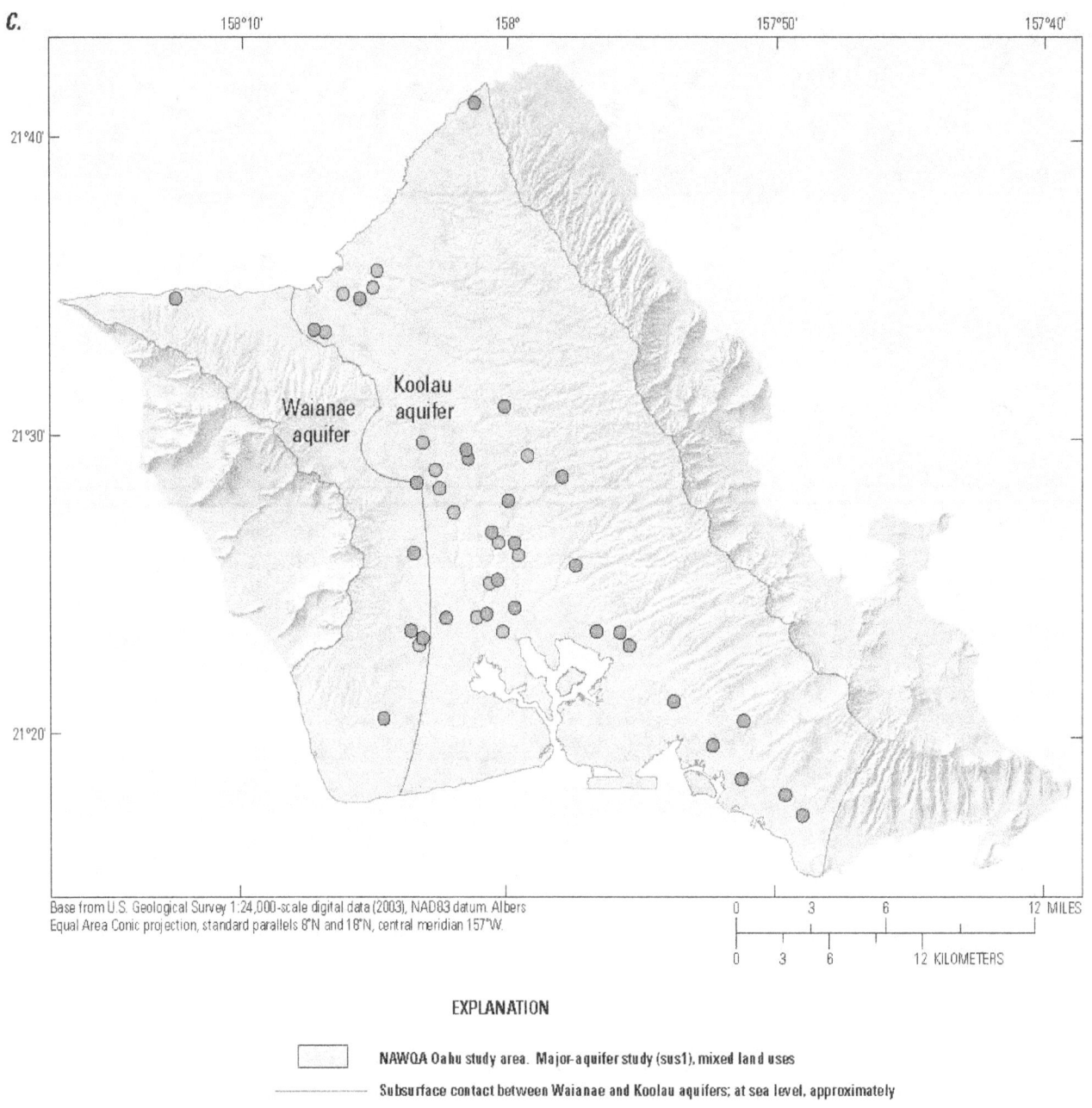

C.

Base from U.S. Geological Survey 1:24,000-scale digital data (2003), NAD83 datum. Albers Equal Area Conic projection, standard parallels 8°N and 18°N, central meridian 157°W.

EXPLANATION

☐ NAWQA Oahu study area. Major-aquifer study (sus1), mixed land uses

─── Subsurface contact between Waianae and Koolau aquifers; at sea level, approximately

◉ Wells tapping major aquifers, public supply (sus1)

◉ Wells tapping major aquifers, shallow monitoring (spcg1)

Figure 12.—Continued.

Table 1. NAWQA groundwater sampling networks in the Columbia Plateau, Snake River Plain, and Oahu principal aquifers.

[Abbreviations and symbols: ccpt, Central Columbia Plateau; usnk, Upper Snake River Basin; oahu, Oahu; lus, land-use study; sus, study-unit (aquifer) survey; DOH, Department of Health; EDB, ethylene dibromide; DBCP, dibromochloropropane; mg/L, milligrams per liter; CFC, chlorofluorocarbon; SF_6, sulfur hexafluoride; *, resampled in NAWQA Cycle II; USGS, U.S. Geological Survey; >, greater than; –, not applicable or data insufficient to compute value]

Network	Study component	What data were collected and why	Sub-network	Area and types of sites sampled	Number of sites	Aquifer lithology	Median well depth (feet)	Median depth to water (feet)	Sampling frequency	Predominant land use or crop type
Central Columbia Plateau, Washington										
ccptsus1* Columbia Plateau	Major-Aquifer Study	Occurrence and distribution of chemicals in major aquifers. Major ions, nutrients, 85 pesticides, 60 volatile organic compounds, dissolved organic carbon, and radon. USGS collected, randomly selected wells.	sus1b*	Public supply wells across Study Unit, randomly selected.	43	Basalt	175	–	Once in 1994; resampled in 2002	Mixed: agriculture, rangeland, and urban
		Estimate risks of pesticide detection. Data included 47 pesticides. DOH collected pesticide kit.	sus1a	Public supply wells across Study Unit, randomly selected.	65	Basalt	200	–	Once in 1994	
			sus1c	Public supply wells across Study Unit, high risk (nitrate > 2 mg/L, shallow).	31	Basalt	185	–	Once in 1994	
ccptlusag1 Agricultural land-use study, Palouse area—dryland grains	Land-use effects—dryland farming	Describe effects of agricultural land use on shallow ground water in the Palouse subunit. Data included major ions, nutrients, 85 pesticides, 60 volatile organic compounds, dissolved organic carbon, and radon. Wells were generally next to fields in the road right-of-way.	lusag1a	Shallow domestic wells in basalt	19	Basalt	102	–	Once in 1993–94	Agriculture (mostly wheat and small grains)
			lusag1b	Very shallow monitoring wells in loess	8	Loess (silt)	37	14	Once in 1993–94	
ccptlusag2* Agricultural land-use study, Quincy-Pasco area	Land-use effects—irrigated row crops	Describe effects of agricultural land use on shallow ground water in the Quincy-Pasco subunit. Data same as above. Wells were generally within 100 feet of row cropped fields.	lusag2a	Shallow domestic wells	30	Sand, gravel, basalt	134	–	Once in 1993–95	Agriculture (mostly potatoes and corn)
			lusag2b*	Very shallow monitoring wells	19	Sand, gravel	38	17	Once in 1993–95; resampled in 2002	

Table 1.   NAWQA groundwater sampling networks in the Columbia Plateau, Snake River Plain, and Oahu principal aquifers.—Continued

[Abbreviations and symbols: ccpt, Central Columbia Plateau; usnk, Upper Snake River Basin; oahu, Oahu; lus, land-use study; sus, study-unit (aquifer) survey; DOH, Department of Health; EDB, ethylene dibromide; DBCP, dibromochloropropane; mg/L, milligrams per liter; CFC, chlorofluorocarbon; SF$_6$, sulfur hexafluoride; *, resampled in NAWQA Cycle II; USGS, U.S. Geological Survey; >, greater than; –, not applicable or data insufficient to compute value]

Network	Study component	What data were collected and why	Sub-network	Area and types of sites sampled	Number of sites	Aquifer lithology	Median well depth (feet)	Median depth to water (feet)	Sampling frequency	Predominant land use or crop type
Central Columbia Plateau, Washington—Continued											
ccptlusor1* Agricultural land-use study, Quincy-Pasco area—orchards	Land-use effects—orchards and vineyards	Describe effects of agricultural land use on shallow ground water in the Quincy-Pasco subunit.	lusor1a	Shallow domestic wells	18	Sand, gravel, basalt	165	55	Once in 1994–95	Agriculture (orchards and vineyards)
		Data same as above. Wells were generally within 100 feet of orchards/vineyards.	lusor1b*	Very shallow monitoring wells	22	Sand, gravel	29	13	Once in 1994–95; resampled in 2002	
ccptspcb1	Special study—Pesticides in ground water at base flow	Nutrients and 47 pesticides to compare nutrient concentrations and pesticide occurrence with what is found in shallow ground water.	—	Smaller irrigation drains (tailwater drainage canals) and subsurface tile-drain outflows	25	Sand, gravel	–	–	Once in winter 1995 and again in winter 1996	Agriculture (row crops and orchards)
ccptspcb2	Special study—EDB synoptic	Fumigants EDB and DBCP at low reporting levels.	—	Shallow domestic wells and irrigation drains	13	Sand, gravel	–	–	Once in 1994	Agriculture (including potatoes in rotation)
Eastern Snake River Plain and Upper Snake River Basin, Idaho and Wyoming											
Major-Aquifer Study / Domestic, irrigation, stock, and public supply wells from a wide range of well depths											
usnksus1 Major-aquifer study		Occurrence and distribution of chemicals in major aquifers. Major ions, nutrients, 87 pesticides, volatile organic compounds, and radon. Data collected in cooperation with the Idaho Statewide Ground-Water Monitoring Program.	—	Snake River Plain, Idaho	43	Basalt	260	–	Once in 1994	Mixed: agriculture and rangeland
usnksus2 Major-aquifer study, Tributary valleys			—	Tributary valleys, Idaho	39	Alluvium	204	–	Once in 1995	Mixed: agriculture and rangeland
usnksus3 Major-aquifer study, Jackson Valley			—	Jackson Valley, Wyoming	20	Alluvium	101	7	Once in 1995	Mixed: forest and rangeland

Table 1. NAWQA groundwater sampling networks in the Columbia Plateau, Snake River Plain, and Oahu principal aquifers.—Continued

[**Abbreviations and symbols:** ccpt, Central Columbia Plateau; usnk, Upper Snake River Basin; oahu, Oahu; lus, land-use study; sus, study-unit (aquifer) survey; DOH, Department of Health; EDB, ethylene dibromide; DBCP, dibromochloropropane; mg/L, milligrams per liter; CFC, chlorofluorocarbon; SF_6, sulfur hexafluoride; *, resampled in NAWQA Cycle II; USGS, U.S. Geological Survey; >, greater than; –, not applicable or data insufficient to compute value]

Network	Study component	What data were collected and why	Sub-network	Area and types of sites sampled	Number of sites	Aquifer lithology	Median well depth (feet)	Median depth to water (feet)	Sampling frequency	Predominant land use or crop type
				Eastern Snake River Plain and Upper Snake River Basin, Idaho and Wyoming—Continued						
				Mostly domestic wells						
usnkluscr1 Agricultural land-use study, Minidoka area	Land-use effects—irrigated row crops	Major ions, nutrients, 87 pesticides, volatile organic compounds, and radon. Data collected in cooperation with the Idaho Statewide Ground-Water Monitoring Program.	–	Minidoka area, mostly surface-water irrigation	29	Sand, gravel	34	8	Once in 1993	Agriculture (mixed crops)
usnkluscr2* Agricultural land-use study, A & B area			–	A&B area, mostly groundwater irrigation	31	Basalt	228	178	Once in 1993; resampled in 2005	Agriculture (mixed crops)
usnkluscr3* Agricultural land-use study, Jerome/Gooding area			–	Jerome/Gooding area, mostly surface-water irrigation	30	Basalt	200	149	Once in 1994; resampled in 2005	Agriculture (mixed crops)
usnkluscr4 Agricultural land-use study, Eden area			–	Eden area, mostly surface-water irrigation	15	Basalt	360	292	Once in 1995	Agriculture (mixed crops)
				Oahu, Hawaii						
oahusus1 Wells tapping major aquifers, public supply	Major-Aquifer Study	Occurrence and distribution of chemicals in major aquifers. Major ions, nutrients, trace elements, 87 pesticides, volatile organic compounds, and radon CFC and tritium-helium dates.	–	Public supply wells across Study Unit, randomly selected.	30	Basalt	512	320	Once in 2000	Mixed: forest, urban, and agriculture (sugarcane, pineapple, mixed crops)
oahuspcg1 Wells tapping major aquifers, shallow monitoring	Special Study—young groundwater in shallow irrigation-recharge layer	Occurrence and distribution of chemicals in irrigation-recharge layer near the water table. Major ions, nutrients, trace elements, 87 pesticides, and volatile organic compounds. CFC, SF_6 and tritium-helium dates.	–	Monitor wells across Study Unit, targeted for shallow open interval just below water table	15	Basalt	693	339	Once in 2001	Mixed: forest, urban, and agriculture (sugarcane, pineapple, mixed crops)

Major-Aquifer Studies

Major-aquifer studies are designed to provide a broad assessment of water-quality conditions of the most important groundwater resources of a geographic area. The large areal and depth dimensions of this groundwater resource require that the study rely primarily on sampling water from existing wells (because the expense required to install monitoring wells was prohibitive). A large proportion of wells in major-aquifer studies are domestic and public-supply wells, so major-aquifer studies can provide valuable insight to the quality of drinking-water resources. Because the major-aquifer studies were so large in extent, they extended over multiple, mixed land uses (unlike land-use studies that targeted a single use such as agricultural).

Major-aquifer studies were conducted in the basaltic-rock aquifers (table 1). Two additional Snake River studies were conducted in basin-fill aquifers of unconsolidated alluvium (Snake River tributary valleys and Jackson Valley, Wyoming). Columbia Plateau and Oahu studies sampled public-supply wells predominantly, whereas the Snake River studies sampled a mix of domestic, irrigation, stock, and public-supply wells to obtain areal coverage of the target aquifers.

Special Studies

Two special groundwater studies were conducted that did not meet the criteria of land-use or major-aquifer studies (table 1). On the Columbia Plateau, a network of domestic wells was sampled for the soil fumigants 1,2-Dibromoethane (ethylene dibromide, EDB), and 1,2-Dibromo-3-chloropropane (DBCP) using low laboratory reporting levels. On Oahu, a network of existing monitoring wells was sampled to study groundwater in the shallow irrigation-recharge layer near the water table.

NAWQA Approach to Groundwater Studies in Each Hydrogeologic Setting

Important aspects of hydrogeology and land use in each area are portrayed on the following block diagrams (figs. 13–15), along with typical well types and configurations. The blocks are not strictly to scale but are meant to span a depth of about 600–1,000 feet on the vertical face of the diagrams.

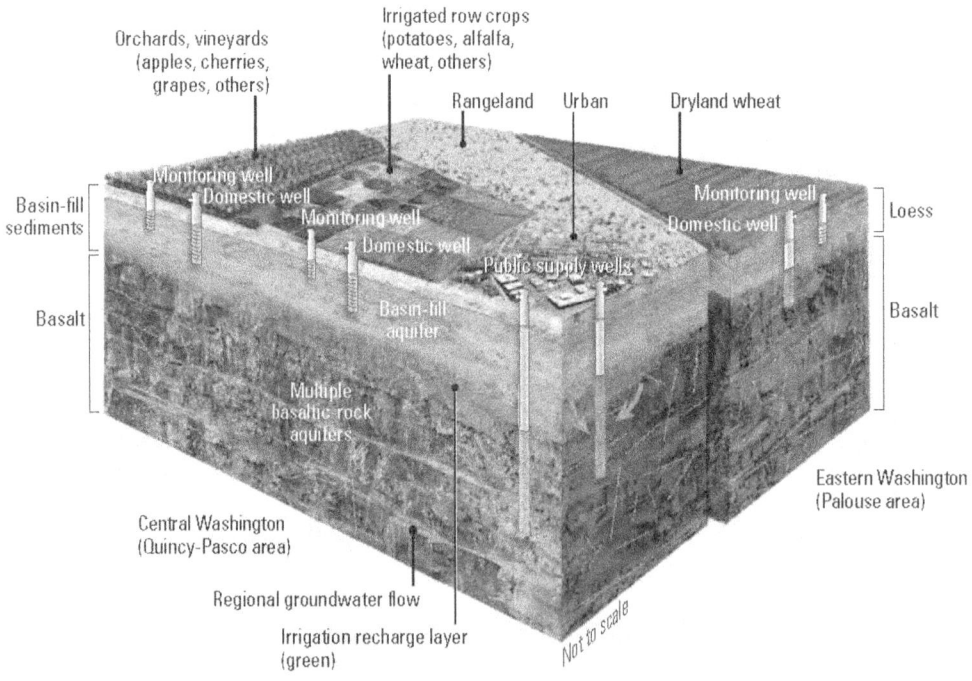

Figure 13.   Block diagram of geologic and hydrogeologic conditions in the Central Columbia Plateau, Washington.

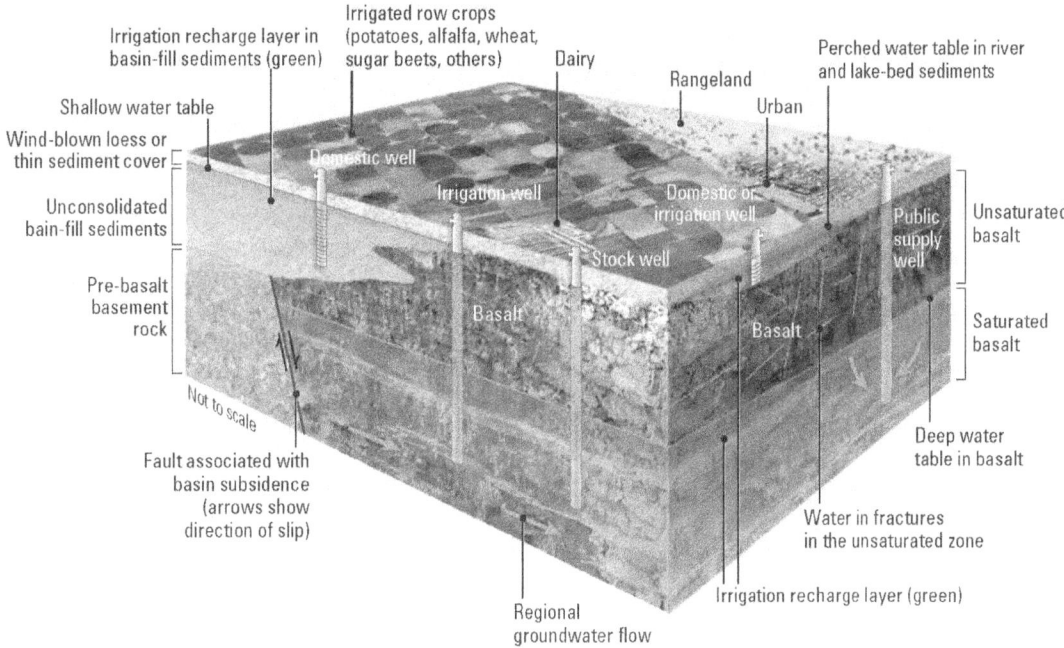

Figure 14. Block diagram of geologic and hydrogeologic conditions in the Eastern Snake River Plain, Idaho.

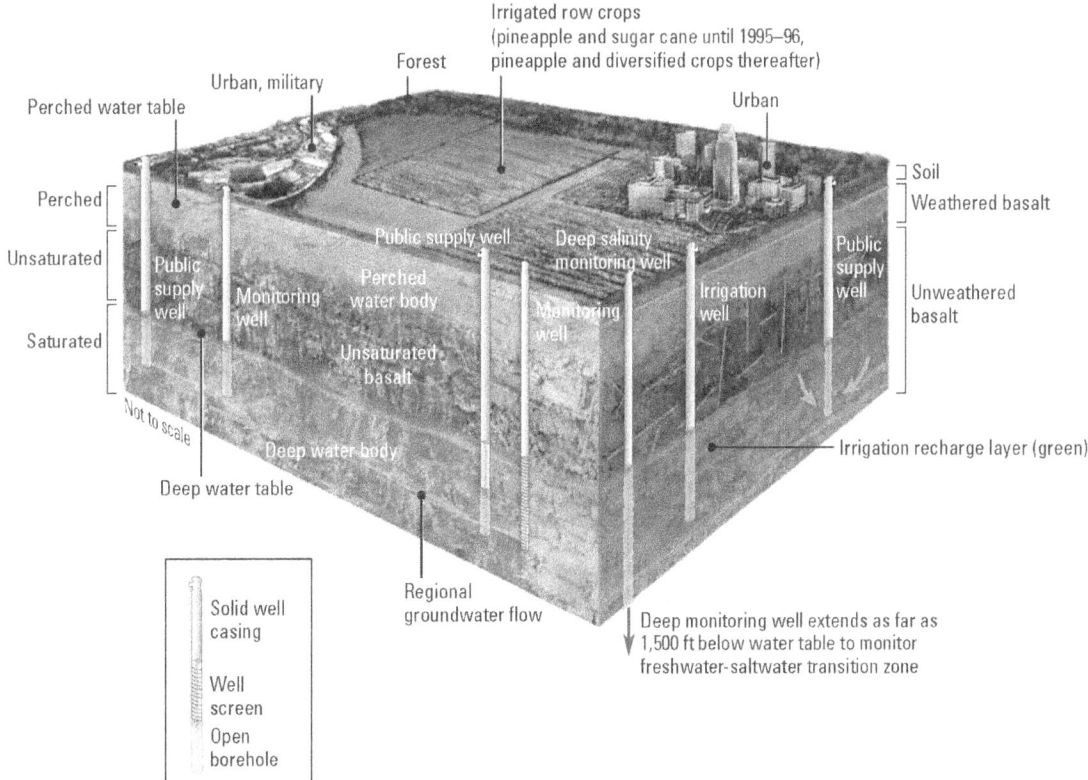

Figure 15. Block diagram of geologic and hydrogeologic conditions in Oahu, Hawaii.

Basin-fill alluvial sediments and loess (windblown silt) blanket the basaltic-rock aquifers throughout much of the Columbia Plateau (fig. 13). Agricultural land-use studies sampled monitoring and domestic wells tapping unconsolidated sediments and basalt. Major-aquifer studies sampled public-supply wells in basalt. Greenish tint in figure 13 identifies the irrigation recharge layer that contains elevated concentrations of nitrate and other contaminants.

The basaltic-rock aquifer occupies much of the Eastern Snake River Plain (fig. 14). Basin-fill sediments encroach from tributary valleys at the side margins of the Plain, and a thin layer of glacial-flood outwash sediments and loess (windblown silt) blanket the upper surface of the basalt. Agricultural land-use studies sampled domestic wells tapping the basalt (3 studies) and shallow perched water in sediments (1 study). Major-aquifer studies sampled domestic, irrigation, stock, and public-supply wells in basalt and unconsolidated basin-fill sediments. Greenish tint in figure 14 identifies the irrigation recharge layer that contains elevated concentrations of nitrate and other contaminants. Depth to the water table in basalt ranges about 50 to 300 feet and depth to the water table in sediments is 10–50 feet in most places.

The basaltic-rock aquifer underlies the entire island of Oahu (fig. 15). Weathering extends to a depth of about 150 feet, and the weathered saprolite contains perched water or near-saturated moisture profiles. Unsaturated, unweathered basalt extends several hundred feet to the deep water table. A major-aquifer study sampled public-supply wells in the basalt aquifer, many of which are solid-cased 50–100 feet into the water to seal off the well against the irrigation recharge layer (greenish tint in figure 15) that contains elevated concentrations of nitrate and other contaminants. A special study sampled existing monitoring wells that were open at shallow depths near the water table to better characterize young water near the deep water table. Depth to the water table in basalt ranges from 100 to 800 feet in most places.

Data Collection

Chemical data used in this report are from samples collected between 1992 and 2010 for the NAWQA land-use studies, major-aquifer studies, and special studies. All three study areas were sampled comprehensively in the first decade of NAWQA (Cycle 1, 1991–2001). A few selected networks were resampled in NAWQA's second decade (Cycle 2, 2001–2012) in the Columbia Plateau and Snake River Plain but not on Oahu, which was discontinued after the first decade. Within the resampled networks, smaller subsets of wells (typically five to seven wells) were also sampled biennially (every 2 years) after the first decadal sampling to characterize temporal variability and trends in water quality that could reasonably be inferred to affect the complete networks.

Samples from the 221 agricultural land-use wells and 271 major-aquifer wells were collected following NAWQA protocols (Koterba and others, 1995). Most data are stored in a central database (NAWQA Data Warehouse;

http://infotrek.er.usgs.gov/apex/f?p=NAWQA:HOME:0) and for this investigation were compiled into Excel spreadsheets. Water samples collected in this study were analyzed for nutrients, major ions, trace metals, radon, pesticides, and volatile organic compounds (VOCs).

The 87 pesticides targeted for investigation were selected on the basis of their agricultural and nonagricultural use, potential environmental significance, and the ability of the USGS National Water Quality Laboratory (NWQL) to quantify them (Larson and others, 1999; U.S. Geological Survey, 1999). The 60 VOCs targeted for analysis were selected on the basis of available information on their occurrence, human and ecological health concerns, ozone depletion potential, use as a fuel additive, and analytical capabilities of the USGS NWQL (Bender and others, 1999).

All samples were analyzed by the USGS NWQL in Denver, Colorado. In circumstances where concentrations of an analyte are sufficiently low to be considered below a detectable quantity, the NWQL reports these concentrations as less than the Laboratory Reporting Limit (LRL). When using the LRL, the risk of reporting a nondetectable analyte concentration when the analyte is actually present (false negative) is less than 1 percent (Childress and others, 1999). The LRL typically is twice the long-term method detection level (LT-MDL). The NWQL LT-MDL is derived by using the standard deviation of at least 24 measurements of spiked matrices containing the analyte(s) of interest. The actual LT-MDL is the minimum analyte concentration(s) measured with 99-percent confidence that the concentration is, in fact, greater than zero. The NWQL reports estimated analyte concentrations when concentrations fall between the LT-MDL and LRL (Childress and others, 1999). Improvements in analytical methods over the years have resulted in changes in LT-MDLs; however, the data used in this report were not recensored to a common detection limit, with the exception of VOCs (see "Volatile Organic Compounds" section later in this report). Estimated concentrations reported by the NWQL were considered detectable concentrations.

Statistical Analyses

Statistical summaries of chemical quality for the NAWQA study units can be biased if water from some wells is sampled more frequently than that from other wells. Where water from wells was sampled multiple times, bias from multiple measurements at the same site was removed by using the oldest analysis (Cycle 1 data). Selecting the oldest analysis is a simple process that is unbiased toward high or low values when there are seasonal or annual trends in the data. More recent samples from Cycle 2 were used in trends analysis. Trends in the datasets were determined using the Wilcoxon-Pratt test (Pratt, 1959). The Wilcoxon-Pratt test is a modification of the Wilcoxon signed rank test so that zero values and tied values can be addressed. Trends were considered significant if the p-value was less than 0.05.

Logistic Regression

Maps showing the probability of atrazine detections (the most commonly detected pesticide) and of elevated concentrations of nitrate in groundwater in the Central Columbia Plateau, Snake River Plain, and Oahu were developed in several steps.

1. All suitable anthropogenic, hydrogeologic, and groundwater quality data were compiled.

2. Groundwater quality data were overlaid with anthropogenic and hydrogeologic data using a geographic information system (GIS) to produce a dataset in which each well had corresponding census, land cover, precipitation, recharge, soils, and well construction data. These data then were downloaded to a statistical software package for analysis.

3. Several preliminary multivariate models with various combinations of independent variables were constructed.

4. The multivariate models that best predict the probability of detecting atrazine and elevated concentrations of nitrate in groundwater were selected.

5. The multivariate models were entered into the GIS, and the probability maps were constructed.

The specific details of data compilation, statistical methods, model development, model validation, and construction of the probability maps are discussed in the following sections.

Compilation of Anthropogenic, Hydrogeologic, and Groundwater Quality Data

Anthropogenic and hydrogeologic data used by this study include census, land cover, precipitation, recharge, soils, and well construction data. Census data for population density were retrieved from the U.S. Bureau of the Census for the 1990 (U.S. Census Bureau, 1991) and 2000 (GeoLytics, 2001) census years. Change of population density was calculated by subtracting year 1990 population density data from year 2000 population density. Mean annual precipitation values in the Central Columbia Plateau and Snake River Plain for the period 1961–90 are from Daly and others (1994, 1997). Mean annual precipitation values in Oahu are from Giambelluca and others (1986). Recharge data in the Central Columbia Plateau were calculated by Hansen and others (1994). Recharge data in Oahu were calculated by Shade and Nichols (1996). Recharge data in the Snake River Plain were calculated by Garabedian (1992). Well construction data include altitude of wellhead, well depth, depth to top of well screen, and depth to bottom of well screen.

Land-cover data in the Central Columbia Plateau and Snake River Plain were compiled from the 1992 national land cover dataset (Vogelmann and others, 2001), which was enhanced with GIRAS land use/land cover data modified

as described in Price and others (2003) to more accurately represent alpine tundra, orchards, vineyards, and residential areas. The land cover data were further processed by calculating the percent of land cover classifications within 500-meter buffers around wellheads in the Central Columbia Plateau and Snake River Plain, which was calculated by Kerie Hitt (U.S. Geological Survey, written commun., 2006). Land cover data in Oahu were mapped by Klasner and Mikami (2003) using digital orthophotoquads photographed during 1998 and 1999. The Klasner and Mikami (2003) data represent late 1990s land use, but do not reflect major changes in land use (urbanization) that have occurred in Oahu during the past 10 years or so. To accurately map historical agricultural lands that only recently have been urbanized, areas used historically for sugarcane and pineapple cultivation mapped by Oki and Brasher (2003, fig. 19) were overlaid with agricultural land-use data mapped by Klasner and Mikami (2003), and agricultural lands were thus delineated. The land cover classifications by Klasner and Mikami (2003) and Oki and Brasher (2003) were modified to correspond to the same land cover classifications used by Vogelmann and others (2001), so that land cover was rated consistently between the Central Columbia Plateau, Snake River Plain, and Oahu. Percents of these land cover classifications within 500-meter buffers around wellheads in Oahu were calculated by this study using methods identical to those used by Kerie Hitt in the Central Columbia Plateau and Snake River Plain.

Soils data were obtained from the State Soil Geographic (STATSGO) database (U.S. Department of Agriculture, 1991). The finer scale Soil Survey Geographic (SSURGO) database (U.S. Department of Agriculture, 1995) was not available for all regions in the Central Columbia Plateau, Oahu, and Snake River Plain. The STATSGO data were not suitable for use by this study in raw form, so STATSGO data compiled by Schwarz and Alexander (1995) were used. These later data included weighted averaging of many of the soil characteristics contained in the database. Soils data in Oahu were not compiled by Schwarz and Alexander (1995), so the STATSGO data in Oahu were compiled by this study using methods identical to those of Schwarz and Alexander (1995). Soils data at the wellhead and within 500-meter buffers around each wellhead were correlated by this study. Soils properties within 500-meter buffers around each well location in the Central Columbia Plateau and Snake River Plain were compiled by Wolock (1997). Wolock (1997) did not compile soils properties within 500-meter buffers around each well location in Oahu, so this study compiled those data using methods identical to Wolock (1997). The U.S. Department of Agriculture (1993) provides more information on these soil characteristics.

Groundwater quality data collected by the U.S. Geological Survey (USGS) National Water-Quality Assessment Program (NAWQA) during 1993–2001 from 205 wells located in the Central Columbia Plateau, 45 wells in Oahu, and 147 wells in the Snake River Plain were used to calibrate the logistic regression models. For wells that had multiple water-quality samples, the first sample collected was used. All data on atrazine and nitrate concentrations in groundwater were converted to binary coding of "zero" for wells with no atrazine detection or nitrate concentration less than 2 mg/L and "one" for wells with atrazine detections or nitrate concentrations greater than or equal to 2 mg/L to satisfy the input data requirements of logistic regression. The breakdown products of deethylatrazine, deethyldeisopropylatrazine, deisopropylatrazine, and 2-hydroxyatrazine data were also evaluated by this study. The presence of atrazine breakdown products indicates the former presence of atrazine, so the groundwater monitoring data were also coded "one" when atrazine breakdown products were detected, even if the parent compound was not present. The Minimum Laboratory Reporting Level for atrazine is 0.001 mg/L.

Statistical Methods and Regression Models

This study used logistic regression (Kleinbaum, 1994; Hosmer and Lemeshow, 2000) to model the probability of detecting atrazine and elevated concentrations of nitrate in groundwater in the Central Columbia Plateau, Snake River Plain, and Oahu. Logistic regression is conceptually similar to multiple linear regression, because relations between one dependent variable and several independent variables are evaluated. In logistic regression, the dependent variable (for this study, atrazine detection or elevated nitrate concentration) was transformed to a binary variable (detection or nondetection). A major advantage of logistic regression over multiple regression is that the former is well suited for analysis of datasets with a large number of nondetections.

Logistic regression calculates several statistical parameters that determine the predictive success of the model. The log-likelihood ratio measures the success of the model as a whole by comparing observed with predicted values (Hosmer and Lemeshow, 2000); specifically, it tests whether model coefficients of the entire model are significantly different from zero. The most significant model is the one with the highest log-likelihood ratio, taking into account the number of independent variables (degrees of freedom) used in the model. The log-likelihood ratio follows a chi-squared distribution, and the computed p-value indicates whether model coefficients are significantly different from zero. In other words, the computed p-value is the significance level attained by the data; the smallest p-value indicates the best model. A p-value of 0.05 indicates a significance level of 95 percent; a p-value of 0.01 indicates a significance level of 99 percent. McFadden's rho-squared (SPSS, Inc., 2000, p. I-571) is a transformation of the log-likelihood statistic and is intended to mimic the r-squared of linear regression. Rho-squared is always between zero and one; a rho-squared

approaching one corresponds to more significant results. Rho-squared tends to be smaller than r-squared, so a small number does not necessarily imply a poor fit. Values between 0.20 and 0.40 indicate good results (SPSS, Inc., 2000, p. I-571).

The partial-likelihood ratio was used to compare nested models to determine the significance of adding one or more new variables (Helsel and Hirsch, 1992; Nolan and Clark, 1997). A nested model contains all of the independent variables in the original model, plus one or more additional independent variables. To determine whether the model is improved by adding the independent variable, the logistic regression model is calculated without that new variable. Logistic regression calculates a partial-likelihood ratio. The logistic regression model then is rerun, this time with the additional new independent variable; the second model also calculates a partial-likelihood ratio. The difference in partial-likelihood ratios between the two models is calculated, and a chi-squared approximation is calculated with degrees of freedom equal to the number of additional variables in the new model. If the p-value from the chi-squared distribution is less than 0.10, the model has been significantly improved at the 90-percent significance level.

In logistic regression, a model is generated that predicts the probability (P) of detecting atrazine or elevated concentrations of nitrate in groundwater similar to equation 1 for each of the two developed models:

$$P = \frac{e^{a+b_1(SU)+b_2(WD)+b_3(L)+b_4(S)}}{1+e^{a+b_1(SU)+b_2(WD)+b_3(L)+b_4(S)}} \tag{1}$$

where

P is the probability of detecting atrazine or elevated nitrate in groundwater,

a is the logistic regression constant,

b_1 is the coefficient for study unit,

SU is the study unit (Central Columbia Plateau, Snake River Plain, and (or) Oahu),

b_2 is the coefficient for well depth,

WD is the well depth,

b_3 is the coefficient for land cover,

L is the land cover (percent agriculture, percent row crops, and (or) percent fallow),

b_4 is the coefficient for soil characteristic, and

S is the soil characteristic (organic mater content and (or) soil permeability).

The coefficients were then standardized following the method of Menard (2002) so that they could be directly compared with one another.

P-values are calculated for each independent variable, which indicate the statistical significance that each variable has on the overall logistic regression model. Independent variables were excluded from the models if their individual p-values were greater than 0.1. The sensitivity is calculated as the number of correctly predicted events (detections) divided by the total number of observed events (SYSTAT Software, 2004). The specificity is calculated as the number of correctly predicted reference events (no detections) divided by the total number of observed reference events. The total correct predictions is calculated as the number of correctly predicted events plus the number of correctly predicted reference events divided by the total number of all events. Sensitivity and specificity data can be used to develop a Receiver Operating Characteristic (ROC) curve, which plots the probability of detecting true signal (sensitivity) and false signal (1−specificity) for the entire range of possible cutpoints (Hosmer and Lemeshow, 2000, p. 160). The area under the ROC curve, which ranges from zero to one, provides a measure of the model's ability to discriminate between those subjects who experienced the outcome of interest versus those who did not. As a general rule, an ROC of about 0.5 suggests no discrimination (equivalent to flipping a coin). An ROC between 0.7 and 0.8 is considered acceptable discrimination. An ROC between 0.8 and 0.9 is considered excellent discrimination. An ROC greater than 0.9 is considered outstanding discrimination, but it would be very rare to observe an ROC this large.

To evaluate model performance, the percentage of actual detections was plotted with the predicted probability of detections by using a deciles of risk calculation, which typically involves partitioning the observations into 10 groups (SYSTAT Software, 2004, p. II–238).

During construction of the logistic regression models, all possible combinations of independent variables were evaluated to develop the most accurate logistic regression models. The models were built by including each individual variable in the model, evaluating the resulting test statistics, and deciding whether to include or reject the variable. Model validity and accuracy were determined by evaluating the log-likelihood ratio, McFadden's rho-squared, the p-values calculated for each independent variable, the model sensitivity and specificity, and the ROC.

Percent land cover, precipitation, population density, soils, and well construction were modeled as continuous variables. Because of their categorical nature, study units were modeled as discrete (design) variables. Discrete variables were coded as "one" if a well was located in a particular study unit and as "zero" if the well was not located in a unit. For example, if a well was located in the Columbia Plateau study unit, then the data base would be coded "Columbia Plateau = 1," "Upper Snake = 0," "Oahu=0." Hosmer and Lemeshow (2000) contain more information on the use of continuous and discrete variables in logistic regression.

Comparison of Concentrations to Human-Health Benchmarks

There are two different human-health benchmarks that are used in this report: Maximum Contaminant Levels (MCLs) and Health-Based Screening Levels (HBSLs). They are used in this report to give perspective on the potential significance of their occurrence in drinking water to human health. For that reason, hereinafter MCLs and HBSLs are referred to together as "human-health benchmarks".

"Maximum Contaminant Levels" are concentrations of constituents established by the U.S. Environmental Protection Agency (USEPA) to help ensure safe drinking water (U.S. Environmental Protection Agency, 2009a). MCLs are legally enforceable concentrations that the USEPA requires public drinking water systems to meet. Public drinking water systems are publicly or privately owned water supply systems that serve at least 25 people or have at least 15 service connections and provide drinking water to the public for at least 60 days per year. Private, individual household wells are not regulated by USEPA, but MCLs provide an initial perspective on the potential significance of contaminant occurrence to human health and can help prioritize further studies. There is currently no MCL for radon in drinking water, although there are two proposed MCLs (U.S. Environmental Protection Agency, 2011d). USEPA has proposed to require community water suppliers to provide water with radon levels no higher than 4,000 picocuries per liter (pCi/L), which contributes about 0.4 pCi/L of radon to the air in your home from showering and other household uses. This requirement assumes that the state is also taking action to reduce radon levels in indoor air by developing USEPA-approved, enhanced state programs for radon in indoor air (called Multimedia Mitigation Programs). Under the proposed regulation, for states that choose not to develop enhanced indoor air programs, individual public drinking water systems will be required to develop their own radon mitigation plan for indoor air or reduce radon levels in drinking water to 300 pCi/L. This amount of radon in drinking water would contribute about 0.03 pCi/L of radon to the air in your home.

Because of health concerns, Hawaii has established lower MCLs for soil fumigants (EDB, DBCP, TCP) than those established by the USEPA. When comparing fumigant concentrations in drinking water to MCLs, this report used the higher USEPA MCLs whenever data from the Columbia Plateau or Snake River Plain were evaluated and used the lower Hawaii MCLs when only Oahu data were evaluated.

MCLs are not established for about two-thirds of the contaminants measured in water by the NAWQA Program and other USGS studies. To supplement existing MCLs, USGS (in collaboration with USEPA and others) established "Health-Based Screening Levels" (HBSLs), which are nonenforceable benchmark concentrations developed using standard USEPA methods and current toxicity information

(Toccalino and others, 2012). HBSLs are equivalent to existing USEPA "Lifetime Health Advisory and Cancer Risk" concentration values (when they exist), except for unregulated compounds for which more recent toxicity information has become available. It is important to note that the presence of contaminants at concentrations greater than benchmarks does not necessarily indicate that adverse effects are certain to occur. Conversely, concentrations that are less than benchmarks do not guarantee that adverse effects will not occur, but they indicate that adverse effects are unlikely.

This report evaluated two concentration levels for human-health benchmarks: the human-health benchmark and 1/10th of the human-health benchmark. Concentrations above 1/10th of the human-health benchmark provide an indication of contaminants that may approach concentrations of potential human-health concern, either individually or as mixtures, and identify those that may warrant additional monitoring and study (DeSimone and others, 2009).

"Secondary Maximum Contaminant Levels" (SMCLs) are nonenforceable guidelines regulating contaminants that may cause cosmetic effects (such as skin or tooth discoloration) or aesthetic effects (such as taste, odor, or color) in drinking water (U.S. Environmental Protection Agency, 2009a). USEPA recommends public drinking water systems meet these secondary standards but does not require public systems to comply. SMCLs are useful guidelines for water from individual household wells.

Water-Quality Conditions

Nutrients

Nitrate concentrations were highest in agricultural areas (fig. 16) and in the shallowest groundwater; however, nitrate contamination exists not only at shallow depths immediately beneath cultivated fields but also in deeper groundwater in the principal regional aquifers used for drinking-water supply. This spatial association results from a combination of applied fertilizers, manure, and irrigation that promotes leaching of the nitrogen through the soil to underlying groundwater. In the Columbia Plateau, nitrate concentrations are generally higher in the southwest, where row crops and orchards are irrigated; nitrate concentrations are generally lower in the eastern area of nonirrigated dryland agriculture. In the Snake River Plain, nitrate concentrations are highest in the western part of the Plain, where cultivation and irrigation are intensive. Nonagricultural lands in the Columbia and Snake areas are mostly rangeland and forest, with relatively small urban areas. On Oahu, much of the central plateau is agricultural,

and elevated nitrate concentrations have resulted from a combination of fertilizer application and irrigation. Most wells with low nitrate concentrations are in forested areas or in the Honolulu urban center to the southeast, where wells draw groundwater that was recharged in upland forests.

The correlation of elevated nitrate with agricultural land use is also evident when results are portrayed by individual well networks (fig. 17). Nitrate concentrations were generally lower in major-aquifer networks and higher in agricultural land-use networks, which were designed to characterize groundwater in agricultural areas where fertilizer use and manure input is known to be intensive. However, nitrate concentrations in major-aquifer networks were not negligible; many concentrations were above background (1 mg/L), and concentrations in several Columbia Plateau public-supply wells (fig. 17) exceeded the drinking-water standard of 10 mg/L (USEPA MCL). These results demonstrate that nitrate has contaminated deep groundwater in the principal regional aquifers used for drinking-water supply. Median nitrate concentrations were similar among major-aquifer networks in basaltic-rock aquifers (fig. 17, left panel, gray bars): about 1 to 1.5 mg/L. However, upper outlier concentrations differed, with concentrations near or exceeding 10 mg/L in the Columbia Plateau and Snake River Plain, whereas maximum nitrate concentration on Oahu was only 5 mg/L. Redox conditions were predominantly oxic in all networks, favoring stability and preservation of nitrate rather than its destruction by denitrification.

Some degree of influence is exerted by well type and depth, irrigation practices, and regional flow patterns in the aquifer systems. Within the agricultural land-use networks, nitrate concentrations were generally higher in unconsolidated basin-fill sediments than in basaltic-rock aquifers. The sediments overlie the basalt aquifers, and wells open to sediments were generally shallower than wells open to basalt (fig. 12, table 1). Higher nitrate in the sediments is an expected result, because agricultural chemicals migrating down from the land surface should be at higher concentrations at shallower depths, unless results are governed more strongly by other factors.

Among agricultural land-use networks in the Columbia Plateau (fig. 17, right panel), nitrate concentrations differed only slightly between domestic-well and monitor-well subnetworks for irrigated row crops and orchards. But the domestic-well subnetwork in the Palouse area of dryland agriculture stands out as an exception, with lower nitrate concentrations likely resulting from a lack of irrigation water that would tend to carry higher nitrate groundwater observed in the shallower monitoring wells to deeper domestic wells in the underlying basalt.

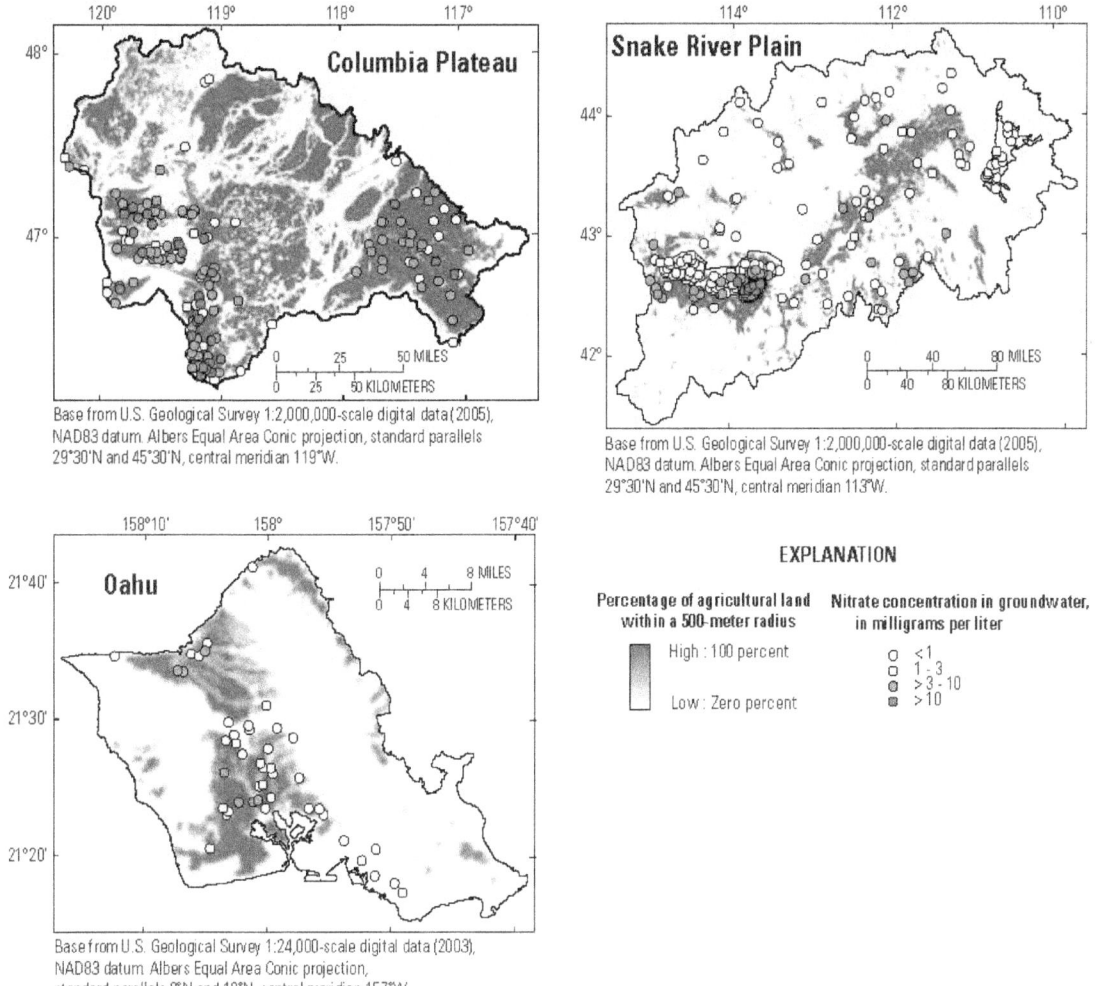

Figure 16. Concentrations of nitrate in groundwater and distribution of agricultural lands in the Columbia Plateau, Snake River Plain, and Oahu study areas.

Agricultural land-use networks in the Snake River Plain consist mostly of domestic wells, so well type can largely be ruled out when explaining differences among networks (fig. 17, center panel, arranged left to right in downgradient direction from east to west). However, well depth, irrigation practices, and regional flow patterns all play a role in influencing nitrate concentrations. The Minidoka network had the highest concentrations. Although the Minidoka area is predominantly irrigated with surface water, groundwater is also used in irrigation. The nitrate concentrations are thought to be high because fertilizer-enriched groundwater is recycled through multiple pumping and irrigation cycles in this shallow perched alluvial aquifer (Rupert, 1997). The other three networks in the Snake River Plain have deeper wells in basalt (fig. 12, table 1), and nitrate concentrations clearly decrease (fig. 17) from network to network in the downgradient direction to the west. This pattern likely reflects upward convergence of better quality regional groundwater as the aquifer thins and the base of the aquifer "shoals" to the west, where the entire flow of the Eastern Snake River Plain aquifer discharges to springs and to the Snake River.

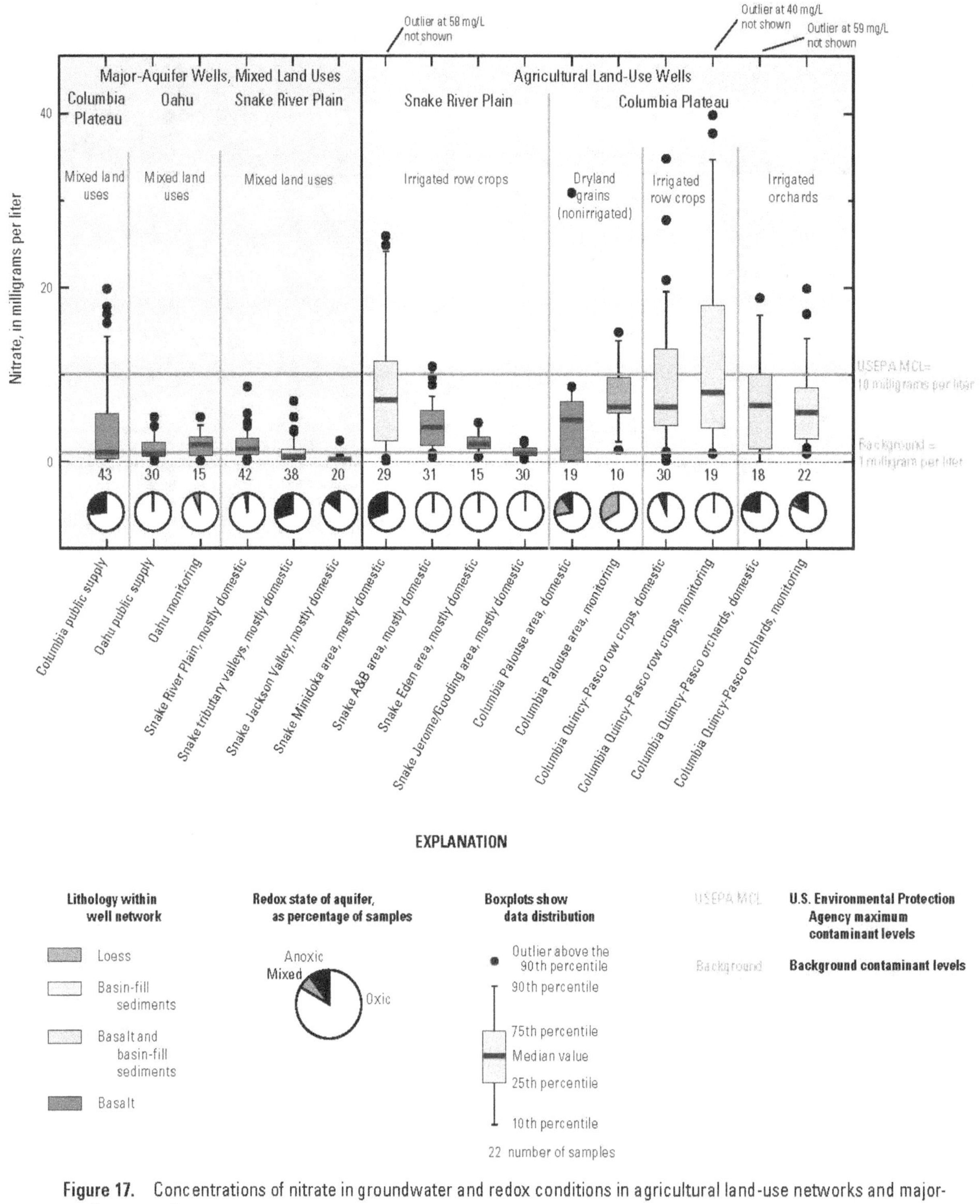

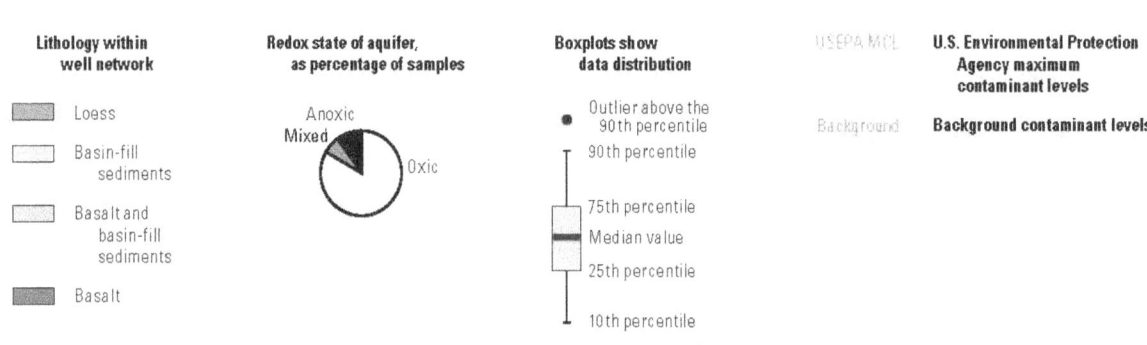

Figure 17.   Concentrations of nitrate in groundwater and redox conditions in agricultural land-use networks and major-aquifer networks.

Nutrient Sources

Mirroring national trends, fertilizer use in the Columbia Plateau and Snake River Plain rose sharply after 1950, leveling off after 1980 in the former but not in the latter (Alexander and Smith, 1990; Battaglin and Goolsby, 1995; Ruddy and others, 2006) (fig. 18). Before World War I, the primary sources of supplemental nitrogen (N) for crops were animal manure, mineral sources such as potassium nitrate, and crop rotation with legume crops such as alfalfa. Synthetic fertilizers were first produced after World War I, when facilities that had produced ammonia and synthetic nitrates for explosives were converted to the production of N-based fertilizers (Rupert, 2008). Inorganic N fertilizer production was small until after World War II, when the production rates increased dramatically (fig. 18). Manure is also used as fertilizer, and its use in the Snake River Plain has increased since 1990 as dairy and beef cattle operations have expanded fivefold. Equivalent fertilizer data are not readily available for Oahu, but the history of fertilizer use likely resembles that for the Columbia Plateau, ramping up after 1950 but leveling off and declining after 1980 as cropland was increasingly converted to suburban residential use. Sugarcane cultivation in Oahu ceased in 1996. Pineapple and diversified crops are still grown, but the southern Oahu pineapple plantation closed in 2008, leaving only the northern plantation.

Although much of the agriculturally applied fertilizer is taken up by crops, a fraction of it leaches through the soil with rain and excess irrigation water and ultimately reaches deep groundwater. Nitrate contamination of groundwater is present in all three study areas as a result of nitrogen leaching. Phosphorus sorbs to soil particles and has much less tendency to leach than nitrogen, but it can be transported to streams with soil runoff. Groundwater beneath cultivated areas did contain elevated phosphorus, but at concentrations typically an order of magnitude lower than nitrate.

Human and animal wastes are also sources of nitrogen and phosphorus. In rural areas, human waste is disposed of in on-site domestic septic systems or cesspools. In urban areas, waste is collected via sewer systems to centralized treatment plants, where solid waste is treated and extracted and later disposed of in landfills or applied as fertilizer biosolids.

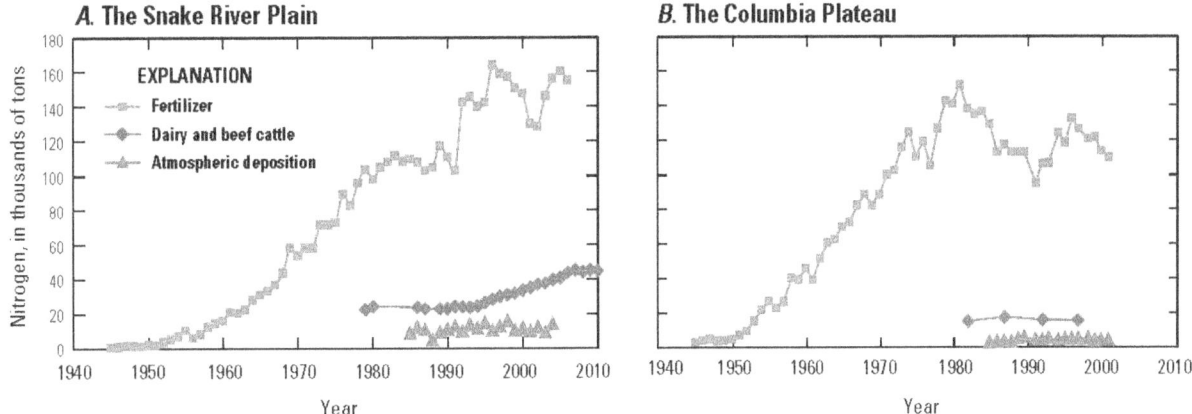

Figure 18.   Nitrogen input over time in (*A*) the Snake River Plain and (*B*) the Columbia Plateau.

The water fraction is treated to environmental standards and reused, applied to the land surface, or disposed of in rivers. Animal waste at larger livestock operations sometimes undergoes treatment, much of it being later used as manure fertilizer. Animal manure is by far the larger source of organic waste in six heavily agricultural counties of the Snake River Plain, where the number of dairy cattle have increased sharply in recent decades (Idaho Agricultural Statistics Service, 1999; U.S. Department of Agriculture, 2010) (fig. 19). The increased manure generated from these cattle has a nitrogen content roughly equivalent to the organic waste that would be produced by an additional 4.6 million to 6.2 million human adults. This is roughly ten times the estimated human population of 491,000 over the entire Snake River Plain aquifer system in 2005.

Trends of Nitrate Concentrations

Nitrate is a good constituent for trend assessment. It is at readily measured concentrations in water from nearly all wells, and variations in anthropogenic sources like fertilizer, manure, and domestic septic systems can be expected to affect nitrate concentrations in groundwater over various time scales. Changes (trends) in concentrations of nitrate can be expected over decades as land-use patterns and agricultural practices shift and evolve.

Assessment of trends is best done with ample data over time, so that many data points establish a pattern, such as a simple rising or falling trend. However, the simplest "trend" that can be determined is a two-point change, with only two data points at time 1 and time 2. This is termed a "change" in this report, whereas "trend" is used only where there are more than two data points.

NAWQA study design includes cycles of sampling entire well networks once every decade for detection of long-term changes and trends. In NAWQA Cycle 2, selected well networks that had been sampled a decade earlier in Cycle 1 were resampled, allowing two-point change detection over the intervening decade (by comparing paired data at time 1 and time 2 at each resampled well). Additionally, a small subset of wells (typically 5 in each network) was sampled biennially (every other year, hence five data points per decade) for interpreting within-decade trends or fluctuations that might be affecting the rest of the network.

Although these NAWQA-collected decadal and biennial data can be used to assess changes and trends in water quality, they span an admittedly short period—roughly a decade and a half from Cycle 1 sampling in 1993–95 to the present year 2011 (sampling on Oahu was a one-time

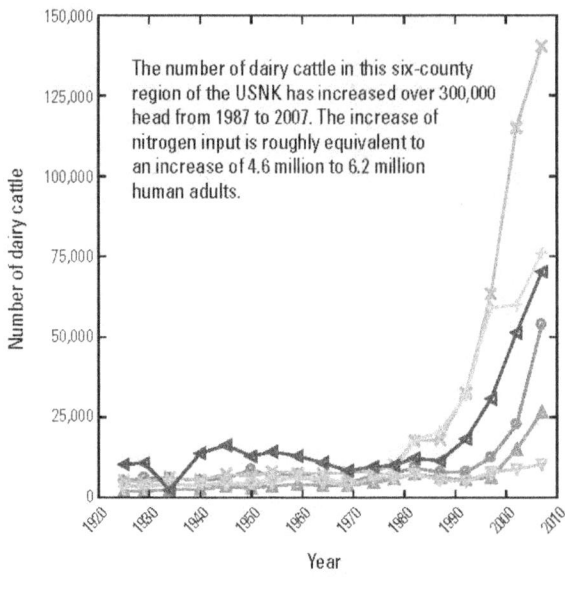

The number of dairy cattle in this six-county region of the USNK has increased over 300,000 head from 1987 to 2007. The increase of nitrogen input is roughly equivalent to an increase of 4.6 million to 6.2 million human adults.

EXPLANATION

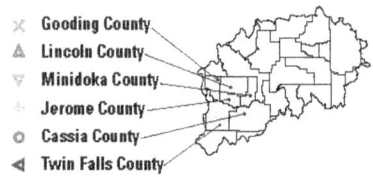

× Gooding County
△ Lincoln County
▽ Minidoka County
+ Jerome County
○ Cassia County
◁ Twin Falls County

Figure 19.   Numbers of dairy cattle in six heavily agricultural counties of the Snake River Plain from 1925 to 2007.

Cycle 1 effort in 2000–2001). Fortunately, longer data sets have been collected by State and local agencies and other USGS programs, providing a means of assessing water-quality trends over longer time scales. These resources are referred to here as "State, local, or cooperative datasets and networks" to distinguish them from NAWQA data. Many of these datasets entailed cooperative funding and effort from multiple agencies.

Decadal Changes in Nitrate Concentration

NAWQA well networks resampled a decade apart showed statistically significant increases in nitrate concentration in the Snake River Plain, but no significant decadal change was detected in Columbia Plateau networks. NAWQA Cycle 1 sampling was conducted in both areas in 1993–95. Cycle 2 resampling was conducted in 2002 in the Columbia Plateau and in 2005 in the Snake River Plain.

The networks were subjected to two-point change detection using graphical and statistical tests that make pairwise comparisons between concentrations at time 1 and time 2 in each well. Individual changes were computed well by well and ranked in the graphs of figure 20, where one can visually judge whether more wells increased or decreased in concentration within a given network. A statistical test (Wilcoxon-Pratt signed-rank statistical test; Pratt, 1959) determined whether any changes in the well network as a whole were statistically significant beyond mere chance occurrence.

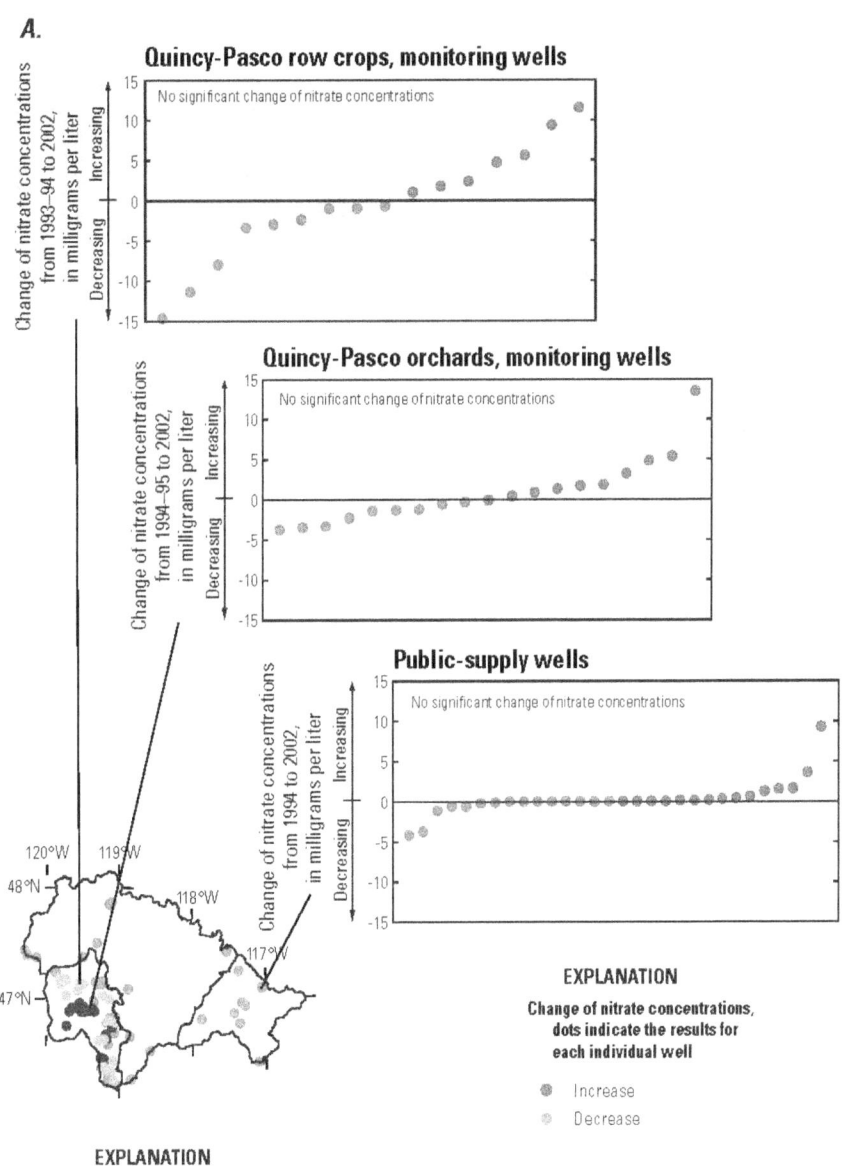

Figure 20. Decadal changes of nitrate concentrations for wells resampled a decade apart (from mid-1990s to 2002) in (*A*) Columbia Plateau and (*B*) Snake River Plain.

In the Snake River Plain networks, more wells increased in concentration (red dots) than decreased (green dots) (fig. 20B). This is apparent visually, and the Wilcoxon-Pratt test confirmed that the increase was statistically significant for each of the Snake networks (Lindsey and Rupert, 2012). In the Columbia Plateau networks, the numbers of red and green dots are roughly equal in each graph, and the Wilcoxon-Pratt test computed no significant change in any of the

Columbia Plateau networks (fig. 20A). The test is conducted on the entire network, and results characterize the network, not individual wells. Although individual wells may show strong increases or decreases in concentration, it is the overall tendency of all wells in the network that determines whether statistically significant change can be detected (or not) for the network overall.

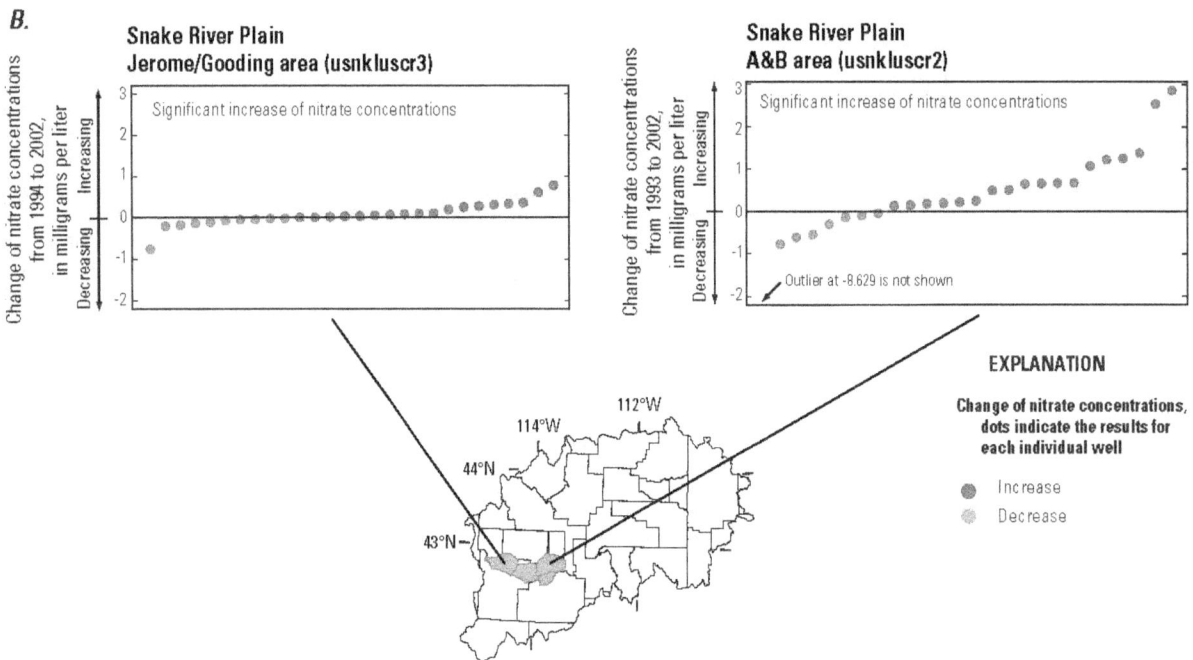

Figure 20.—Continued.

Long-Term Nitrate Trends

Datasets with numerous samples over time were analyzed for trends in nitrate. The datasets were collected by a variety of Federal, State, and local agencies, including the USGS. For many wells, no statistically significant trends could be detected. However, a number of wells in each of the study areas showed significant trends in nitrate, and representative examples are discussed below.

In the Snake River Plain, wells and springs sampled by USGS projects and the State of Idaho showed mostly increasing trends in nitrate concentration (fig. 21). Box Canyon Springs showed a steady increase since the 1970s (fig. 21), and Niagara and Clear Springs have shown significant increases in nitrate through the 1990s (Idaho Department of Environmental Quality, 2000). In addition, groundwater sampled by the U.S. Bureau of Reclamation showed mostly increasing nitrate concentrations since 1980 (Rupert, 1997).

In the Columbia Plateau, NAWQA biennial data showed mostly decreasing nitrate trends (fig. 22), although one example of an increasing trend is shown at upper left (site 471449119522801). Additionally, a previous USGS study with a 474-well dataset in the same Columbia Plateau study area found statistically significant decreases in nitrate concentrations at some wells from 1998 to 2002 (Frans and Helsel, 2005). No trend was found in the 474-well dataset as a whole, but nitrate concentrations decreased in "high-nitrate wells" (wells with nitrate concentrations greater than 10 mg/L). A previous study of those 474 wells, but using only samples collected during 1998 (Frans, 2000), indicated an increased probability of nitrate exceeding 10 mg/L in wells (1) that had shallow well casings, (2) were in areas of high fertilizer application, or (3) were located on soils with high infiltration rates. These factors helped to identify wells that were most likely to have decreasing nitrate concentrations in response to improvements in irrigation and fertilizer practices. The observed nitrate declines in "high-nitrate wells" likely reflect concerted efforts toward better practices by stakeholders who have long been aware of the problem of high nitrate in groundwater, including farmers, agricultural extension services, and State and County agencies.

On Oahu, annual nitrate data from the Hawaii Department of Health allowed trends assessment from 2000 to 2010, the decade following NAWQA Cycle 1 sampling in

2000–2001. Of the several public-supply wells shown in south Oahu (fig. 23) most have slight but statistically significant downward trend in nitrate concentrations (graphs *A-D* have a significant trend, and the Hoaeae wells, graphs *E* and *F*, have no significant trend).

The graphs illustrate an important basinwide spatial pattern of nitrate increasing in the downgradient direction (the direction of groundwater flow), from rainy mountains in the northeast corner of the map and then following the general direction of stream valleys to discharge points at wells and springs near Pearl Harbor. Groundwater at a well in the upland forest (fig. 23, graph *C*) has a "background" nitrate concentration of about 0.45 mg/L. That groundwater flows southwest beneath agricultural croplands, where large volumes of irrigation recharge have "lain in" on top of the regional flow, creating a distinct upper layer of warmer groundwater containing elevated nitrate, pesticides, and other constituents. This "irrigation-recharge layer" is roughly 100 to 150 feet thick in south Oahu and has been studied in detail since the 1960s (Visher and Mink, 1964; Tenorio and others, 1969). Wells drawing from the layer of degraded water currently have nitrate concentrations as high as 5 mg/L (graphs *A*, *B*, *E*). Figure 23 (graph *G*) illustrates depth stratification of the irrigation recharge with better quality water underneath; nitrate concentrations in the shallower well have averaged nearly twice the nitrate concentration in the deeper well over the decade 2000–2010 (4.6 vs 2.4 mg/L).

It is somewhat surprising that downward trends in nitrate are not steeper in south Oahu, given recent changes in land use and irrigation practices. This may be due, in part, to residential fertilizer use as large tracts of former agricultural land have been converted to suburban use beginning in the mid-1960s and are continuing today (the road network shows current suburban tracts over former cultivated areas). Furrow (or "field flood") irrigation of sugarcane was notoriously inefficient, with as much as 50 percent of applied water lost to deep groundwater as nitrate-laden irrigation recharge. This was replaced with more efficient (10-percent loss) drip irrigation by about 1980, dramatically lessening irrigation recharge. However, the thick accumulated layer of irrigation recharge in the aquifer has persisted through the intervening three decades, 1980 to 2010, which illustrates the long "flushing time" or "response time" of the hydrologic flow system within the regional aquifer.

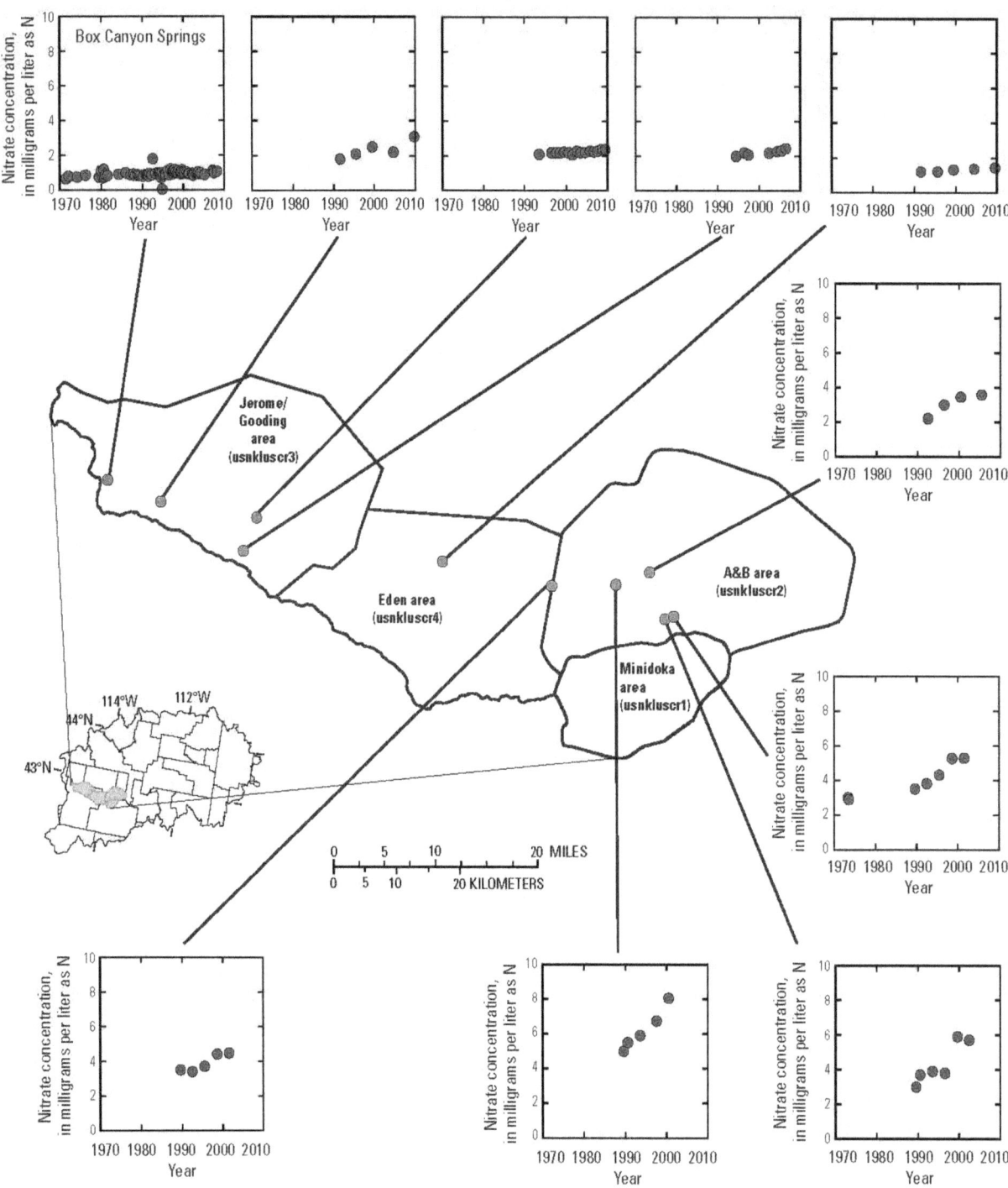

Figure 21.   Trends in groundwater nitrate concentrations in the Snake River Plain agricultural areas from 1970 through 2010. Data were collected by a variety of groundwater-monitoring projects, and only sites with statistically significant trends are shown.

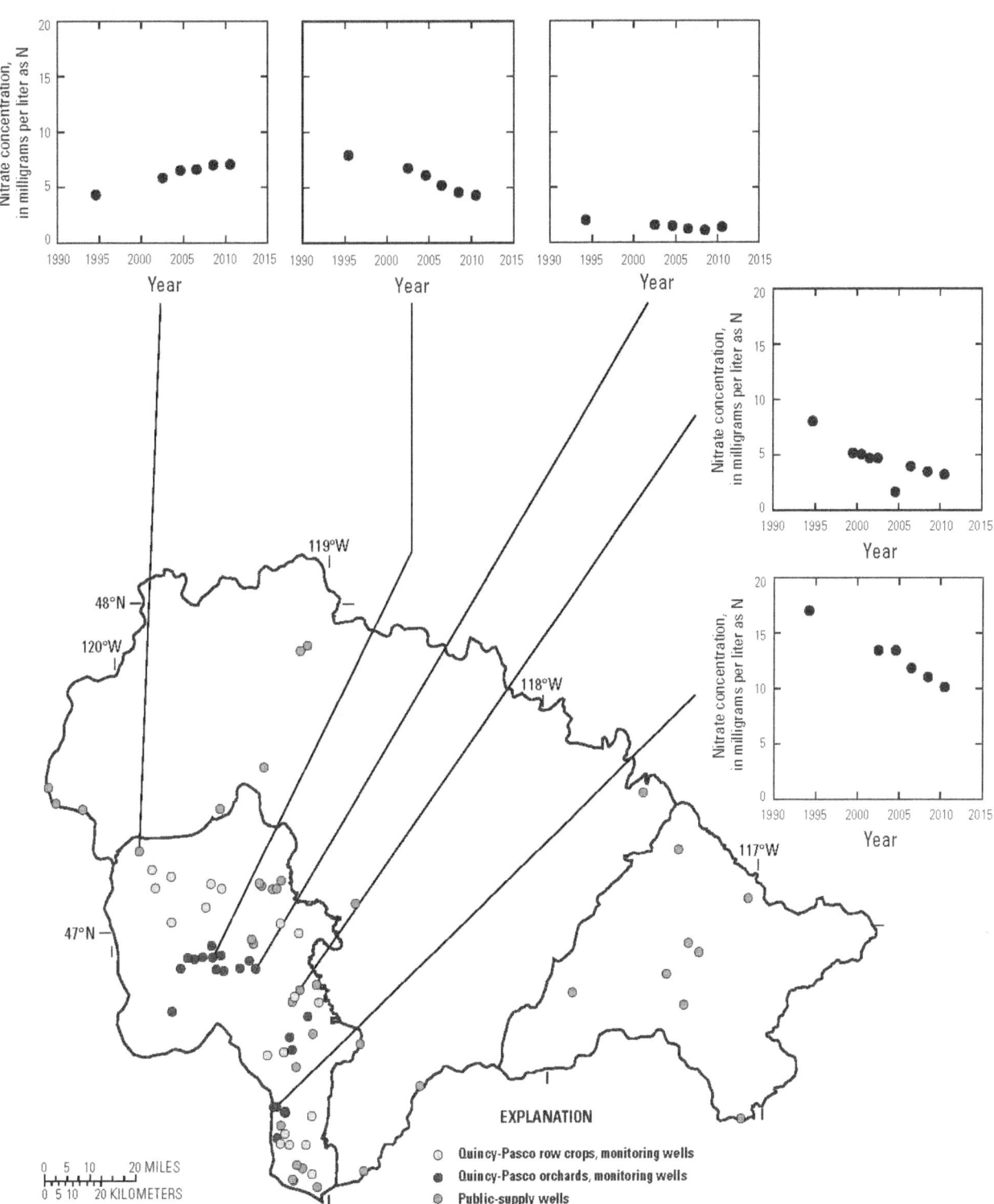

Figure 22. Trends in groundwater nitrate concentrations in selected wells in the Columbia Plateau since the early 1990s. Data are NAWQA decadal and biennial data, and only sites with statistically significant trends are shown.

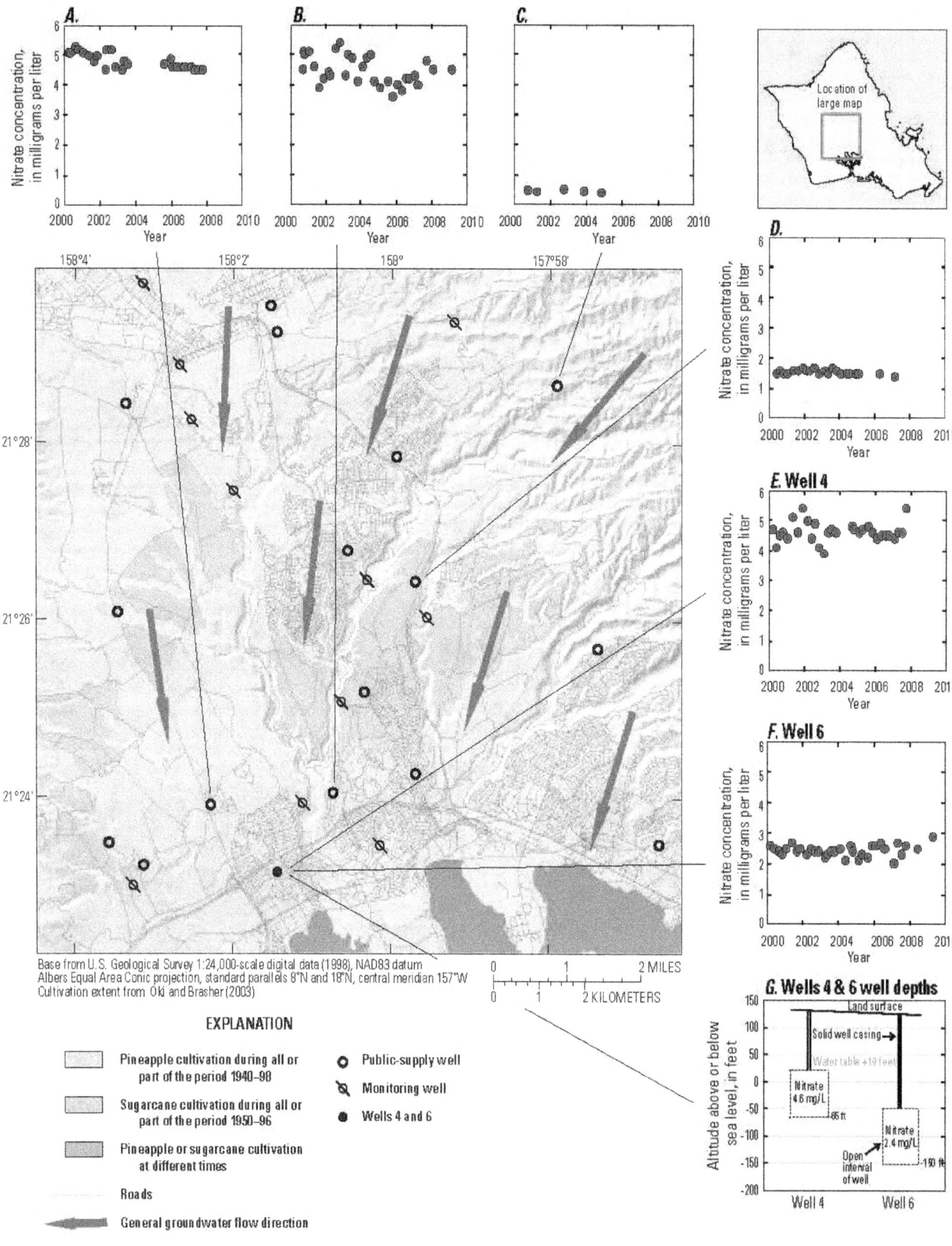

Figure 23.   Nitrate trends in south Oahu public-supply wells since 2000.

Predicting the Distribution of Elevated Nitrate Concentrations

A logistic regression model was developed to predict the probability of detecting elevated concentrations of nitrate. The binary response variable was defined by dividing the nitrate concentrations into those greater than or equal to 2 mg/L and those that were less than 2 mg/L. The threshold of 2 mg/L was chosen because background nitrate concentrations are generally less than 2 mg/L (Nolan and others, 1998) and because of the much lower concentrations of nitrate in the Oahu study area. A higher threshold would have resulted in virtually all of the Oahu wells being coded as nonexceedances.

Study unit, well depth, percentage of agricultural land within 500 meters of the well, and organic-matter content of the soil were significant variables in the nitrate model (table 2). The p-values of each variable incorporated in the nitrate model were all less than 0.0008, with the exception of the Central Columbia Plateau study unit variable, and several were less than 0.0001.

The positive and negative signs of the model coefficients were consistent with expectations. Well depth showed a negative correlation with elevated nitrate concentrations, as nitrate concentrations tend to decrease with depth below land surface. The relation between percentage of agricultural land use near a well and elevated nitrate concentrations was positive because the primary source of nitrate in the study areas is agricultural fertilizer. The relation between elevated nitrate concentrations and organic-matter content of the soils was negative. Soils with high organic-matter content may have an increased likelihood of fostering denitrifying conditions, thereby decreasing the amount of nitrate available to leach into the groundwater. The standardized coefficients indicate that the percentage of agriculture near the well has the greatest impact on nitrate concentrations, followed by the well depth.

Overall performance of the nitrate model was good, with the chi-squared p-value calculated from the log likelihood ratio of the entire model less than 0.0001 and a McFadden's rho-squared of 0.188. The ROC was greater than 0.78. To help confirm that the nitrate models are calibrated to the groundwater quality data, regressions were made between the percentage of actual detections of nitrate and the predicted probability of detecting elevated nitrate (fig. 24).

Table 2. Logistic regression coefficients and individual p-values of independent variables significantly related with the detection of nitrate concentrations greater than or equal to 2 milligrams per liter in groundwater in the Central Columbia Plateau, Upper Snake River Plain, and Oahu study units.

[**Abbreviation:** m, meter; < less than]

Independent variable	Unstandardized nitrate model regression coefficients	p-value	Standardized coefficients
Logistic regression constant	0.8288	0.2029	–
Study unit—Central Columbia Plateau	–0.3716	0.4628	–0.21407
Study unit—Upper Snake River Plain	–1.7793	0.0008	–0.20909
Depth of well	–0.00277	<0.0001	–0.26753
Percent agricultural within 500-m radius	0.0237	<0.0001	0.27394
Organic matter content of soil	–1.1142	0.0001	–0.20351

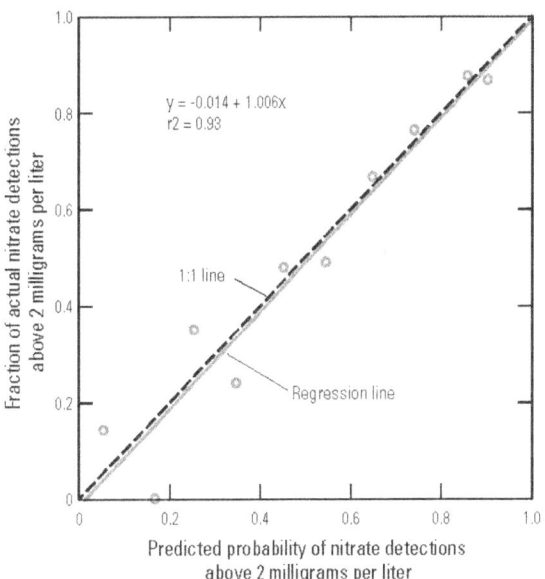

Figure 24. Percentage of actual detections of nitrate in groundwater greater than 2 milligrams per liter versus the predicted probability derived from the nitrate logistic regression model of detections of nitrate in groundwater greater than 2 milligrams per liter.

The percentage of predicted detections of elevated nitrate was determined by dividing the predicted probabilities for the study areas into deciles, or groupings of 10 percent (0 to 10 percent, greater than 10 to 20 percent, greater than 20 to 30 percent, and so on). The percentage of elevated nitrate detections within each group then was calculated and included in the regressions shown in figure 24. The nitrate model exhibits good calibration, with an r-squared value of 0.93.

Maps showing the probability of detecting nitrate concentrations greater than or equal to 2 mg/L (fig. 25) in the Columbia Plateau, Snake River Plain, and Oahu were constructed using the logistic regression models. Before constructing the maps, all GIS data were converted to grids. Data in Oahu were converted from polygon coverages to 30-meter grids. All GIS data in Central Columbia Plateau and Snake River Plain were converted to 1-kilometer grids because the percent land cover grids (Naomi Nakagaki, U.S. Geological Survey, written commun., 2006) were already mapped at 1-kilometer. Then, the logistic regression models similar to equation 1 were entered into a GIS and a probability rating was calculated for each of the grid nodes in each of the three study units.

Areas with a high percentage of land in crops (such as potatoes or sugarcane) (fig. 25), and soils with low amounts of organic matter, are most likely to have elevated nitrate concentrations in the groundwater. Areas where agricultural activities were absent had much lower probabilities of detecting elevated nitrate concentration. The Columbia Plateau had a very high probability of having elevated nitrate concentrations, with most of the land area having greater than a 50-percent probability of elevated nitrate concentrations. Oahu and the Snake River Plain had a much lower probability of having elevated nitrate concentrations because of their lower percentage of agricultural land.

Surface-Water Receptors of Groundwater Discharge

Groundwater that is not captured by wells and used consumptively eventually discharges from the regional aquifer systems. In the Columbia Plateau and Snake River Plain, discharge is by visible springs in cliffs and bluffs or by invisible seepage to stream channels and marshy riparian zones along the streams. On Oahu, aquifer discharge is to the coastal zone, in bays and estuaries, along beaches, and to streams where stream channels intersect the water table. In Oahu, there are no major rivers approaching the size of the Columbia or Snake Rivers.

Excess nutrients in discharging groundwater and in surface-water runoff can foster eutrophication in these receiving water bodies. Eutrophication is the excessive growth of aquatic plants or periodic blooms of microalgae that can emit toxins or that die, settle to the bottom, and decay—depleting dissolved oxygen and causing fish kills. Excess phosphorus and excess nitrate can both lead to eutrophication. Phosphorus is mainly transported to receiving waters with suspended sediment (sorbed to soil particles in runoff). Nitrate is mainly delivered in dissolved form, in discharging groundwater within the three areas. Where groundwater discharges to river bottoms and riparian zones, high organic-matter content in sediments generally causes reducing, anoxic conditions in which some nitrate is broken down by denitrification, sending nitrogen gas back to the atmosphere. However, groundwater discharging from valley-wall springs and bluffs tends to remain oxic, with little loss of nitrate to denitrification. This occurs along some sedimentary bluffs along the Columbia River but is perhaps best exemplified by the numerous large valley-wall springs in the Thousand Springs area of Idaho, which is the terminal discharge zone for the entire unused groundwater flow from the Eastern Snake River Plain aquifer.

Pesticides in Groundwater of the Columbia Plateau, Snake River Plain, and Oahu

Atrazine and its degradate (a compound produced from the breakdown of a parent pesticide), deethylatrazine, were the most commonly detected pesticides in groundwater sampled in the Columbia Plateau and Snake River Plain (fig. 26). Bromacil was the most commonly detected pesticide on Oahu. The other pesticides most commonly detected in the study areas include simazine, hexazinone, metribuzin, diuron, prometon, metolachlor, p,p'-DDE, dieldrin, 2-4-D, and alachlor. Gilliom and others (2006) reported that many of the same pesticides have been detected in groundwater in many regions of the United States. Atrazine has also been widely detected in Europe's groundwater, where it was banned for use in all of the European Union's member states in 2005 because of concerns of groundwater contamination (Ackerman, 2007, p. 446). Of the 11 most commonly detected pesticides in groundwater of the Columbia Plateau, Snake River Plain, and Oahu, 8 of them are herbicides. Soil fumigants, a specialized type of pesticide widely used in these areas to control soil-borne pests such as fungi, nematodes, and weeds, are discussed in a separate section of this report.

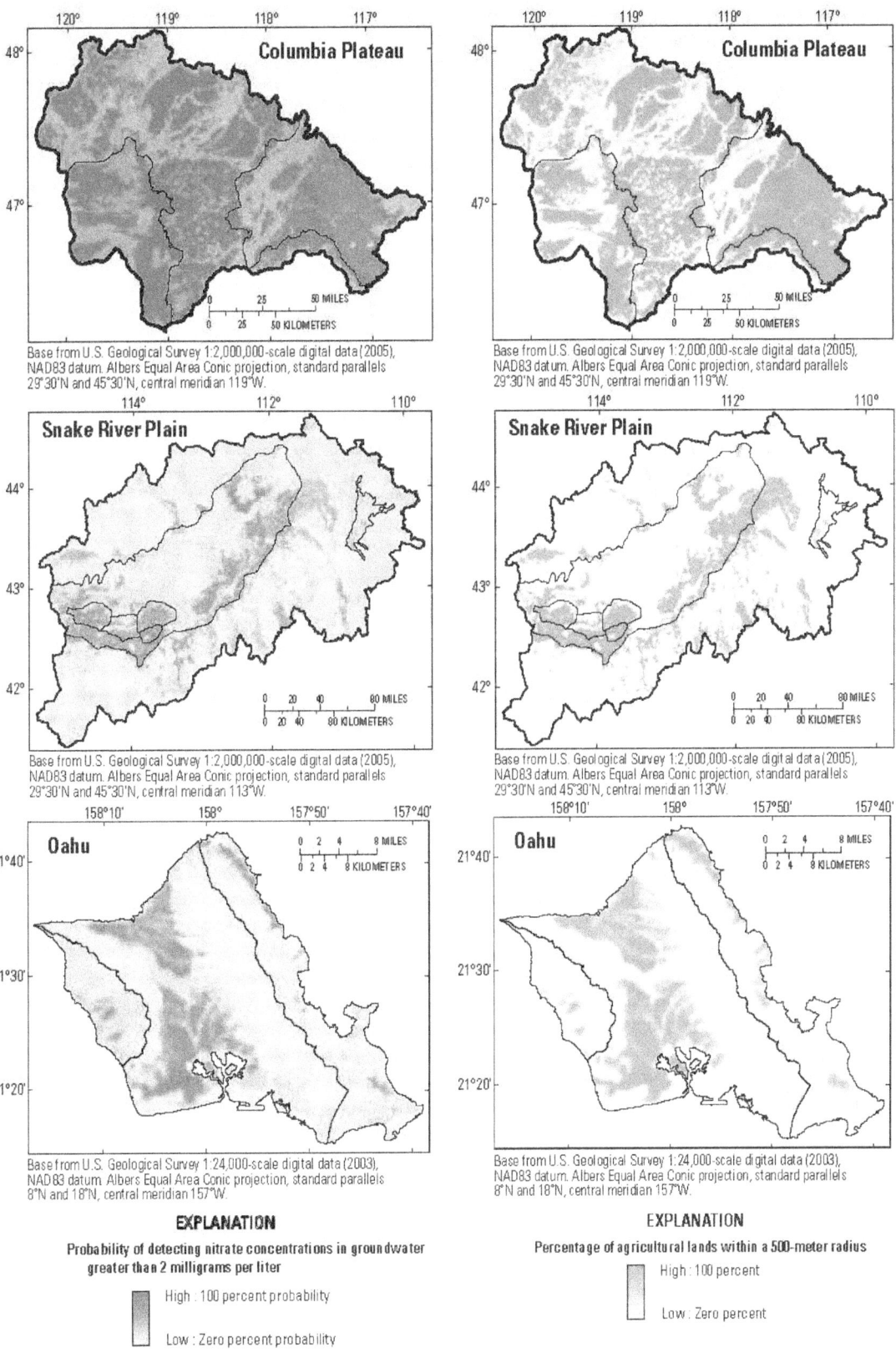

Figure 25.   Probability derived from the nitrate logistic regression model that nitrate will exceed 2 mg/L in groundwater of the Snake River Plain, Columbia Plateau, and Oahu and extent of agricultural land use in those areas.

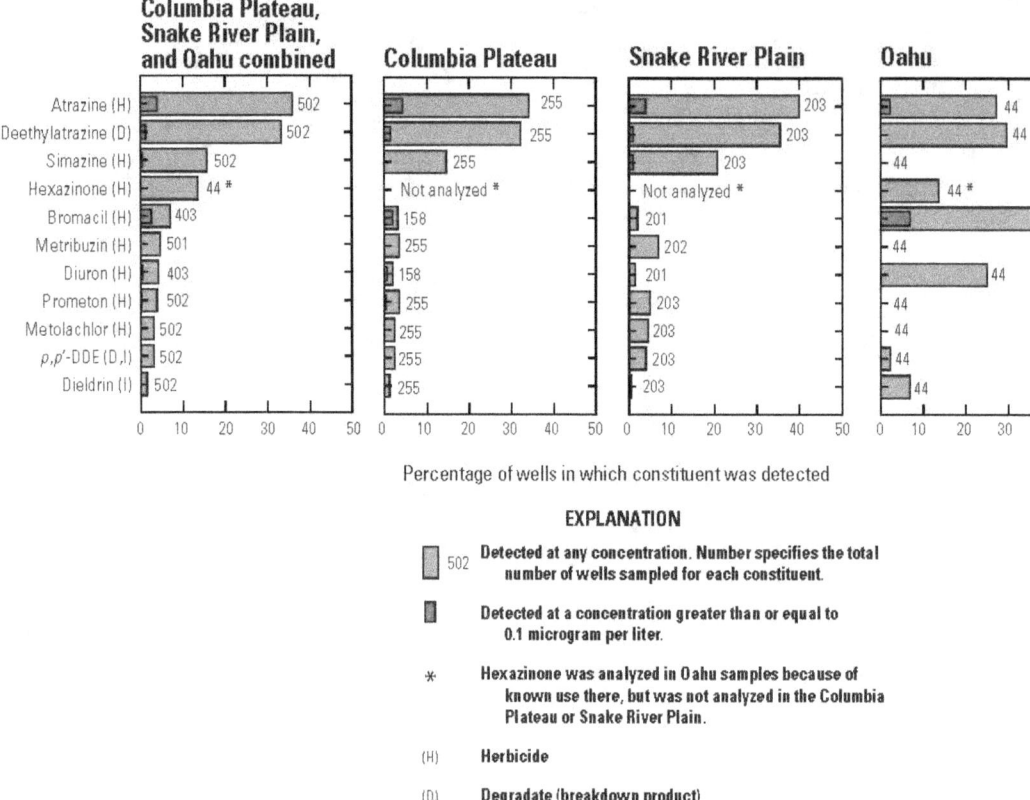

Figure 26. Commonly detected pesticide compounds in water from wells in the Columbia Plateau, Snake River Plain, and Oahu.

Although detections of pesticides in groundwater were widespread in the Columbia Plateau, Snake River Plain, and Oahu, concentrations were generally below human-health benchmarks. For example, the human-health benchmark for atrazine in drinking water is 3 µg/L (U.S. Environmental Protection Agency, 2009a), but the largest concentration observed was less than 1 µg/L. The human-health benchmark for simazine is 4 µg/L, and the largest concentration observed in this report was 0.14 µg/L. The human-health benchmark for hexazinone is 400 µg/L (Toccalino and others, 2012); the largest concentration of hexazinone was about 0.09 µg/L. The human-health benchmark for bromacil is 70 µg/L (Toccalino and others, 2012); the largest concentration measured in groundwater for this study was 14 µg/L.

The differences in pesticide detections in groundwater of the Columbia Plateau, Snake River Plain, and Oahu reflect different crops and pesticides favored in those areas. Some pesticides were detected in groundwater samples from all three

study areas, but other pesticides were detected only in samples from Oahu, or only in samples from the Columbia Plateau and Snake River Plain (fig. 26). This is because some pesticides are broad-spectrum pesticides that are used on many crops in different areas of the United States. Other pesticides are used on specific crops, so they are detected only in groundwater underlying those particular crops. For example, atrazine is one of the most widely used herbicides in the United States, and it is used for weed control on many of the major row crops grown in the Columbia Plateau and Snake River Plain (U.S. Environmental Protection Agency, 2011c). Atrazine is also used for weed control in sugarcane and pineapple, and is used to a lesser extent on residential lawns (U.S. Environmental Protection Agency, 2011c). Because atrazine is used on many different crops in the Columbia Plateau, Snake River Plain, and Oahu, atrazine and its degradate, deethylatrazine, are some of the most frequently detected pesticides in groundwater of all three study areas (fig. 26).

Simazine, metribuzin, metolachlor, and prometon were detected in groundwater of the Columbia Plateau and the Snake River Plain, but were not detected in groundwater from Oahu (fig. 26). This is because simazine, metribuzin, and metolachlor are used on the row crops grown in the Columbia Plateau and Snake River Plain (such as potatoes, barley, and alfalfa), but not on any major crops grown on Oahu (U.S. Environmental Protection Agency, 1995, 2003b). Sometimes, pesticides are combined together for more effective weed control. For example, simazine and metribuzin can be combined with atrazine for more effective weed control in row crops (U.S. Environmental Protection Agency, 2011b).

Hexazinone, bromacil, and diuron were frequently detected in groundwater from Oahu, but were detected much less frequently in groundwater from the Columbia Plateau or the Snake River Plain (fig. 26). This is because hexazinone, bromacil, and diuron are used for weed control in sugarcane and pineapple crops, historically the major cash crops on Oahu (Pesticide Management Education Program, 1993; U.S. Environmental Protection Agency, 1996). Bromacil and hexazinone have been used to control weeds in pineapple fields in central Oahu since the 1970s and 1980s, respectively (Zhu and Li, 2002). Bromacil and diuron were detected at much lower percentages in groundwater from the Columbia Plateau and the Snake River Plain, probably because their primary use in those areas is to control weeds and brush in nonagricultural areas, rather than on major crops. These nonagricultural areas include utility right-of-ways, ditch banks, railroads, electrical switching stations, and industrial yards (U.S. Environmental Protection Agency, 1996; U.S. Environmental Protection Agency, 2003a).

The insecticides DDT and dieldrin are long banned but persistent pesticides that are still being detected in groundwater of the Columbia Plateau, Snake River Plain, and Oahu. DDE was detected in groundwater samples from the Columbia Plateau, Snake River Plain, and Oahu at low concentrations (about 0.001 µg/L). DDE can originate by breakdown of DDT or as a contaminant in commercial DDT formulations (U.S. Department of Health and Human Services, 2002). Similar to DDT, DDE can persist for a long time in the environment. Most DDT in soil is broken down to DDE and DDD (degradates of DDT) by microorganisms, and the half-life of DDT can be as much as 15 years (U.S. Department of Health and Human Services, 2002). For many years, DDT was one of the most widely used pesticides in the United States (U.S. Environmental Protection Agency, 1975). DDT was first synthesized in 1874, but its effectiveness as a pesticide was not discovered until 1939. During World War II and afterward, the United States produced large quantities of DDT for control of vector-borne diseases such as typhus

and malaria abroad. DDT was banned for agricultural use in the United States in 1972, but before its cancellation, approximately 1.35 billion pounds of DDT was used in the United States (U.S. Environmental Protection Agency, 1975). Although DDE is believed to be relatively immobile in groundwater (U.S. Department of Health and Human Services, 2002), it was still detected at trace concentrations in groundwater sampled from all three areas studied for this report.

Dieldrin is an insecticide that was originally developed in the 1940s as an alternative to DDT (U.S. Environmental Protection Agency, 2003c). Dieldrin was manufactured as a parent compound but also results from degradation of aldrin, another insecticide (Brasher and Anthony, 2000). Under most environmental conditions, aldrin is largely converted to dieldrin, which is significantly more persistent (U.S. Environmental Protection Agency, 2003c). Dieldrin proved to be a highly effective insecticide and was very widely used during the 1950s to early 1970s as a broad-spectrum soil insecticide for the protection of various food crops, as seed dressings, to control infestations of pests like ants and termites, and to control several insect vectors of disease. In 1972, the EPA cancelled all but three specific uses of these compounds (subsurface termite control, dipping of nonfood plant roots and tops, and completely contained moth-proofing in manufacturing processes), and these were by 1987 voluntarily cancelled by the manufacturer (U.S. Environmental Protection Agency, 2003c). Dieldrin does not easily break down over time and tends to biomagnify (which means concentrations tend to increase in tissue as it is passed up the food chain) (U.S. Environmental Protection Agency, 2003c). Dieldrin was used extensively for termite control on Oahu (Brasher and Anthony, 2000), which may help explain the higher occurrence in groundwater of Oahu than of the Columbia Plateau or the Snake River Plain.

Although concentrations were low, pesticides were commonly detected in water from domestic and public-supply wells. Pesticides were detected in water from 30 percent or more of domestic wells sampled in the Columbia Plateau and Snake River Plain (fig. 27) (this study did not sample domestic wells in Oahu). Atrazine was the most commonly detected pesticide in domestic wells, followed by deethylatrazine and simazine. Most homeowners do not routinely test water from their domestic wells, probably because of the costs of pesticide analyses and because they are unaware of the common occurrence of pesticides in water from domestic wells. Most domestic wells in the Columbia Plateau and Snake River Plain are located in rural agricultural areas where pesticide use is greatest.

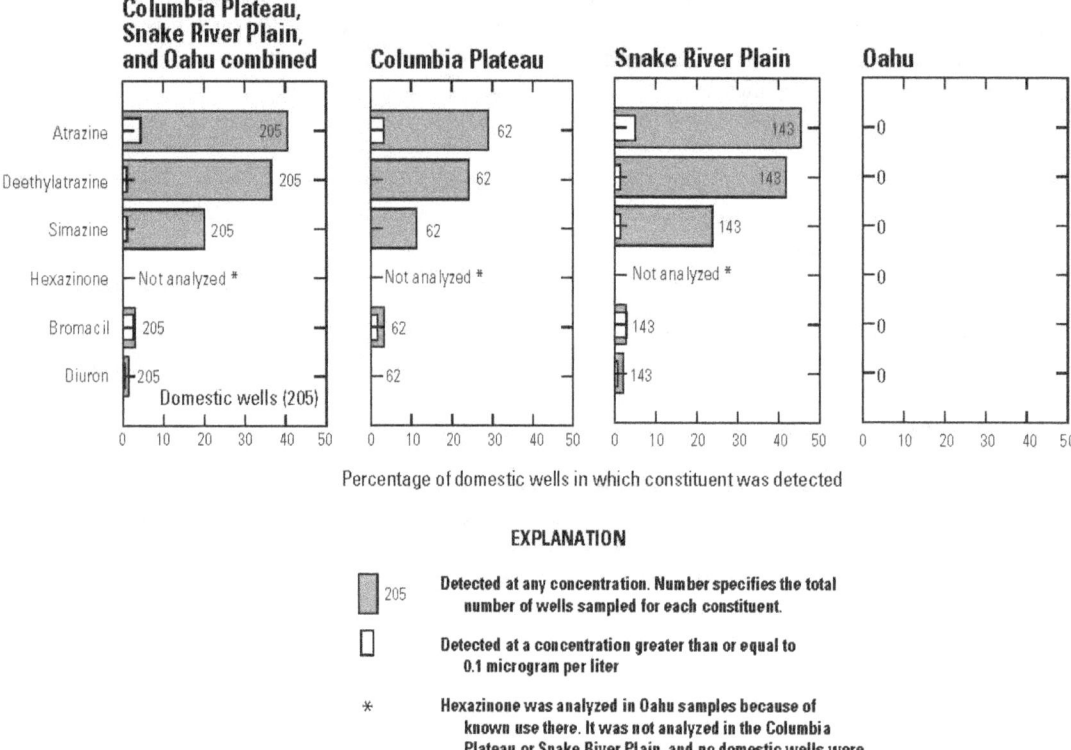

Figure 27.   Pesticides detected in domestic wells in the Columbia Plateau and the Snake River Plain (domestic wells in Oahu were not sampled).

Agricultural areas had the greatest occurrence of pesticides in groundwater in the Columbia Plateau, Snake River Plain, and Oahu (fig. 28). This is not surprising, because that is where pesticide use is greatest and because agricultural land-use studies incorporated carefully selected well networks to characterize groundwater directly beneath cultivated land. However, pesticides were also detected in a large percentage of samples from major aquifer studies, where well networks encompassed mixed land uses over larger expanses of the regional aquifers. This indicates that pesticide contamination can be quite widespread and of regional extent within the deep, principal aquifers used for public water supplies (most wells in the major-aquifer studies were public-supply wells). To some extent it may also reflect pesticide contamination from a variety of land uses and not just from agriculture.

Pesticides were also commonly detected in water from public-supply wells in the Columbia Plateau and Oahu, but hardly at all in the Snake River Plain (fig. 29). Although concentrations were low compared to human-health benchmarks, several pesticides were detected in as many as 20 to 40 percent of the public-supply wells sampled in the

Columbia Plateau and Oahu. Public-supply wells are routinely monitored for pesticides by the water supply providers. If pesticide concentrations exceed human-health benchmarks, the wells are often decommissioned or expensive water treatment systems are installed.

Water samples from 50 percent of the wells sampled in all three study areas contained one or more pesticides, with some samples containing as many as 10 pesticides (fig. 30). Water samples containing two or more pesticides were much more common than samples containing just one pesticide, meaning that pesticides most commonly occur in groundwater as mixtures. This is consistent with what was observed nationally; Gilliom and others (2006) reported that 47 percent of wells sampled in agricultural areas and 37 percent of wells in urban areas had detections of two or more pesticides or degradates. Deethylatrazine, a degradate of atrazine, was included in this compilation, but other pesticide degradates such as 3,4-dichloroanaline (a degradate of diuron) were not included because laboratory methods for these degradates were not yet developed when groundwater samples were collected in the Columbia Plateau and Snake River Plain.

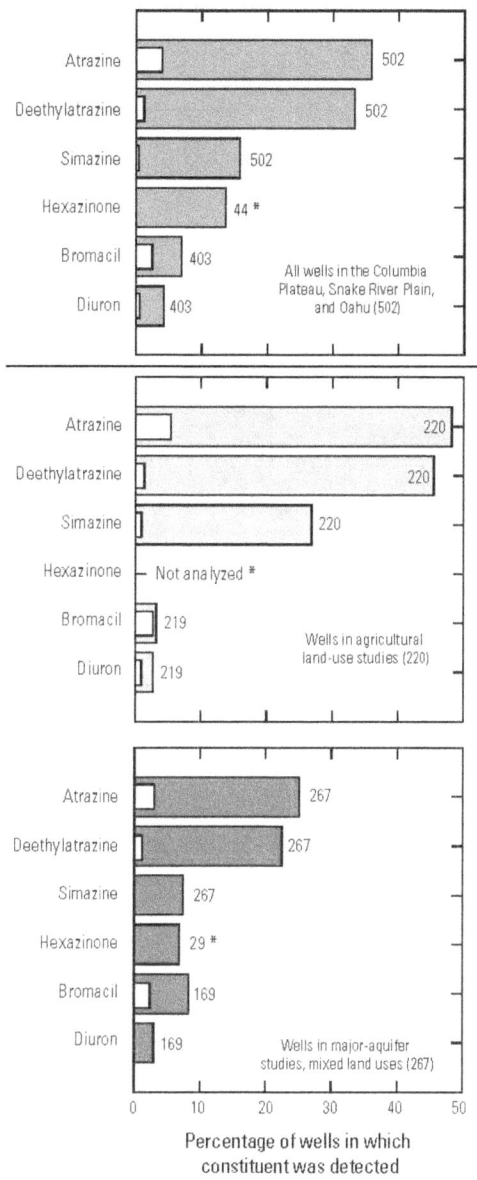

Percentage of wells in which
constituent was detected

EXPLANATION

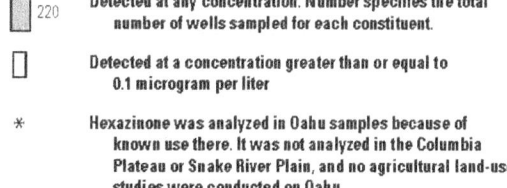

220   Detected at any concentration. Number specifies the total
number of wells sampled for each constituent.

Detected at a concentration greater than or equal to
0.1 microgram per liter

*   Hexazinone was analyzed in Oahu samples because of
known use there. It was not analyzed in the Columbia
Plateau or Snake River Plain, and no agricultural land-use
studies were conducted on Oahu.

Figure 28.   Pesticides detected in land-use and major
aquifer studies in the Columbia Plateau, Snake River
Plain, and Oahu.

Fumigants and VOCs were not included in this compilation of multiple detections because they are discussed in later sections of this report. Combining fumigant and VOC detections with pesticide detections in the same water sample would have increased the total number of constituents detected in water from many wells.

Mixtures of pesticide compounds in groundwater are of particular interest for human-health reasons. Toxicologists have evaluated health risks and established drinking-water regulations for many single compounds but risks associated with compound mixtures are far less known and may be greater than those of single compounds in some cases (Bartsch and others, 1998; Carpenter and others, 1998).

Atrazine and its degradate, deethylatrazine, constituted the most common mixture observed (fig. 31). It is common to see the parent compound and its degradate in the same groundwater sample. However, atrazine and deethylatrazine were also co-detected with other pesticides, and these mixtures can be indicative of the region where they were applied. For example, simazine and metolachlor are commonly applied in combination with atrazine (U.S. Environmental Protection Agency, 2011c). Atrazine and simazine or metolachlor form some of the more common mixtures of pesticides observed in groundwater (fig. 31), but they were only detected in groundwater from the Columbia Plateau and the Snake River Plain. This is because simazine and metolachlor are not used on Oahu (U.S. Environmental Protection Agency, 2011c). As another example, atrazine used to be applied to pineapple (U.S. Environmental Protection Agency, 2003d), and bromacil and diuron are also applied to pineapple (Pesticide Management Education Program, 1993). Atrazine, deethylatrazine, bromacil, and diuron formed the fourth most common mixture of pesticides observed, but only in Oahu (fig. 31). One possible explanation for the widespread occurrence of pesticide mixtures is that some pesticides are applied to crops as mixtures to begin with, or in successive years. Mixed land use and the mixing of contaminant plumes in groundwater may also create mixtures of pesticide compounds that were not applied together at the land surface.

The pesticide mixtures observed in the Columbia Plateau and the Snake River Plain were consistent with those observed nationally, for which Gilliom and others (2006) reported that more than 30 percent of all unique mixtures found in streams and groundwater in agricultural and urban areas contained the herbicides atrazine (and deethylatrazine), metolachlor, simazine, and prometon. Mixtures of atrazine, deethylatrazine, bromacil, and diuron were unique to Oahu, where crops not normally grown in the continental United States are grown (sugarcane and pineapple).

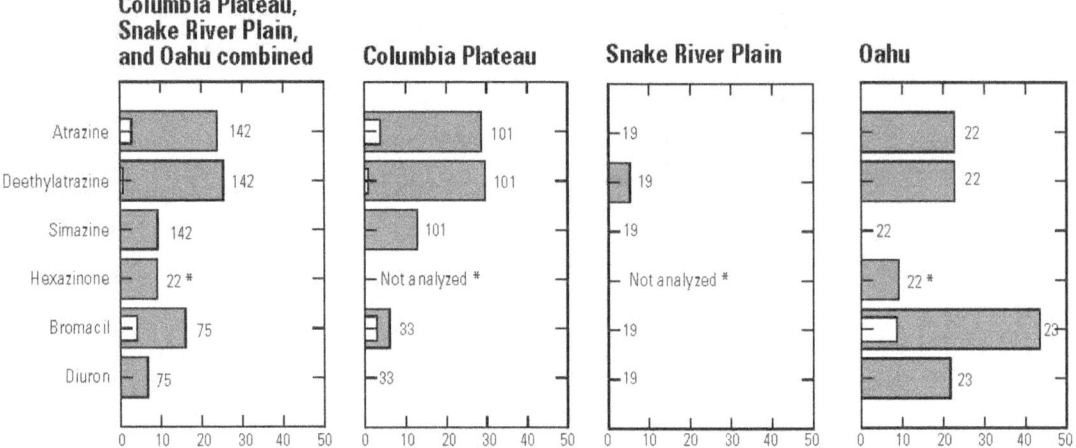

Percentage of public supply wells in which constituent was detected

EXPLANATION

142   Detected at any concentration. Number specifies the total number of wells sampled for each constituent.

Detected at a concentration greater than or equal to 0.1 microgram per liter

*   Hexazinone was analyzed in Oahu samples because of known use there. It was not analyzed in the Columbia Plateau or Snake River Plain

Figure 29.   Pesticides detected in public-supply wells in the Columbia Plateau, Snake River Plain, and Oahu.

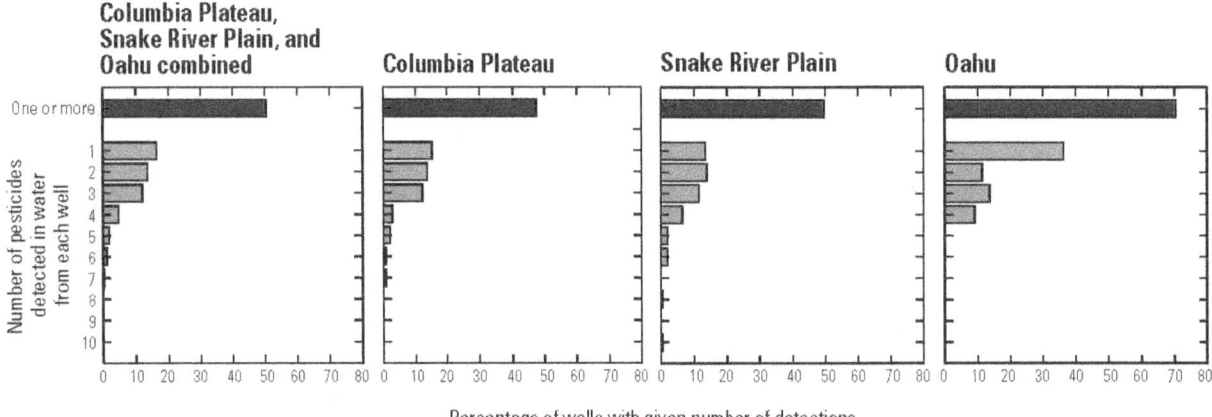

Percentage of wells with given number of detections

Figure 30.   Percentage of wells with multiple detections of pesticides in the Columbia Plateau, Snake River Plain, and Oahu.

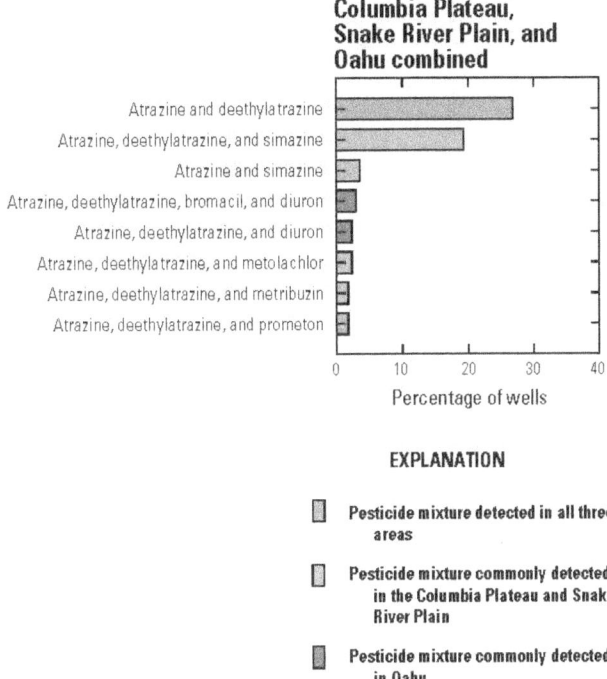

Columbia Plateau, Snake River Plain, and Oahu combined

EXPLANATION

☐ Pesticide mixture detected in all three areas

☐ Pesticide mixture commonly detected in the Columbia Plateau and Snake River Plain

☐ Pesticide mixture commonly detected in Oahu

Figure 31. Percentage of wells in the Columbia Plateau, Snake River Plain, and Oahu with detections of the most common pesticide mixtures.

Predicting the Distribution of Atrazine Detections

A logistic regression model was developed to predict the probability of detecting atrazine (the most commonly detected pesticide) at a concentration of 0.001 µg/L, which is the minimum laboratory reporting limit for atrazine, in groundwater in the Central Columbia Plateau, Snake River Plain, and Oahu. Deethylatrazine, deethyldeisopropylatrazine, deisopropylatrazine, and 2-hydroxyatrazine data were combined with the atrazine data because they are breakdown products of atrazine, and their detection indicates the former presence of atrazine. The groundwater quality data were transformed to a binary response variable of detect or no detect before logistic regression modeling.

Study unit, well depth, percent of row-crop land within a 500-meter buffer around the well head, percent of fallow land within a 500-meter buffer around the well head, soil organic-matter content, and soil permeability were significant variables in the model (table 3). The individual p-values of the significant variables in the atrazine model were all less than 0.06, and most were less than 0.04.

The positive and negative signs of the model coefficients were consistent with expectations. Well depth showed a negative correlation with atrazine occurrence, indicating that as well depth increases, atrazine occurrence decreases. There was a positive correlation with percent of row crops, which is the agricultural land where most atrazine is expected to be applied. There was a negative correlation with fallow agricultural lands. Fallow agricultural lands make up a very small percentage of land cover in the Central Columbia Plateau, Snake River Plain, and Oahu, but they are not expected to receive atrazine applications. Soil organic matter and soil permeability had negative correlations and positive correlations, respectively, with atrazine occurrence. This is consistent with what is expected—atrazine tends to sorb onto soils with high organic matter contents, and soils with greater permeability tend to have more infiltration. The standardized coefficients indicate that the depth of the well has the greatest impact on atrazine detections, followed by the organic matter of the soil.

Several preliminary models using various combinations of independent variables were developed before selecting the final atrazine model shown in table 3. Depth to top of well screen was a significant variable in several preliminary models, as were population density and change of population density. Precipitation was a significant variable is several preliminary models, but when probability maps were produced for Oahu it became apparent that precipitation was not a viable variable because precipitation had an anomalously large influence in the probability ratings, resulting in lowland areas having anomalously large probability ratings.

Table 3. Logistic regression coefficients and individual p-values of independent variables significantly related with the detection of atrazine in groundwater in the Central Columbia Plateau, Upper Snake River Plain, and Oahu study units.

[**Abbreviation:** m, meter; <, less than]

Independent variable	Unstandardized atrazine model regression coefficients	p-value	Standardized coefficients
Logistic regression constant	0.1648	0.7222	–
Study unit—Upper Snake River Plain	0.0642	0.0405	0.11777
Study unit—Oahu	1.135	0.0534	0.14022
Depth of well	–0.00318	<0.0001	–0.30081
Percent row crops within 500 m radius	0.0152	0.0064	0.15933
Percent fallow within 500 m radius	–0.0243	0.0555	–0.12043
Organic matter content of soil	–1.1688	0.0072	–0.20867
Permeability of soil	0.0853	0.0394	0.11534

Subsequent logistic regression models that did not incorporate precipitation were much more effective in Oahu, and they still produced effective models in the Central Columbia Plateau and the Snake River Plain. Percent of urban lands was a significant variable in some preliminary models, which is to be expected because atrazine is used on lawns and recreational grasses. These preliminary models were not selected for the final model because they had inferior statistical performance.

Soils properties at point locations had more significant statistical correlations with atrazine detections than soils properties mapped within 500-meter buffers around the wellheads. This is probably because the STATSGO soils data were generalized from 1:24,000-scale to 1:250,000-scale maps to construct the STATSGO soils database. Averaging the soils data within 500-meter buffers only serves to further generalize the data, reducing statistical significance.

Recharge was never a significant variable in any of the models, probably because other variables, such as crop type and organic matter content, are more closely related to atrazine occurrence. Atrazine use data are available for Central Columbia Plateau and Snake River Plain but are not available for Oahu, so atrazine use was not included in the models. It is likely that atrazine use would be a significant independent variable, if such data were available for Oahu.

Percent row crops was the only agricultural land-use variable positively related with atrazine occurrence, whereas nitrate was positively correlated with all agricultural lands (orchards, pasture/hay, row crops, small grains, and fallow). This is probably because atrazine is only applied to irrigated row crops, but nitrogen fertilizers are applied to all agricultural lands.

Overall performance of the atrazine model was good with the chi-squared p-value calculated from the log likelihood ratio of the entire model less than 0.001 and a McFadden's rho-squared of 0.184. To help confirm that the atrazine models are calibrated to the groundwater quality data, regressions were made between the percentage of actual detections of atrazine and its breakdown products and the predicted probability of detecting atrazine and its breakdown products (fig. 32). The percentage of predicted detections of atrazine and its breakdown products was determined by dividing the predicted probabilities for the study areas into deciles, or groupings of 10 percent (0 to 10 percent, greater than 10 to 20 percent, greater than 20 to 30 percent, and so on). The percentage of atrazine detections within each group then was calculated and included in the regressions shown in figure 32. The atrazine model exhibits good calibration, with an r-squared value of 0.97.

Maps showing the probability of detecting atrazine at or greater than concentrations of 0.001 µg/L (fig. 33) in the Central Columbia Plateau, Snake River Plain, and Oahu

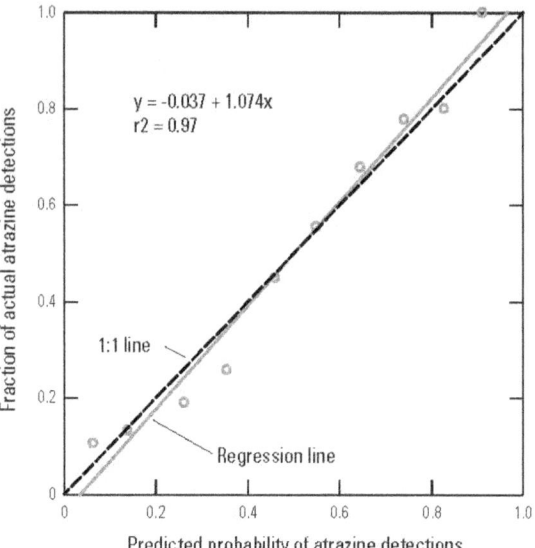

Figure 32.   Percentage of actual detections of atrazine in groundwater versus the atrazine logistic regression model predicted probability of detections of atrazine in groundwater.

were constructed using the logistic regression models. Before constructing the maps, all GIS data were converted to grids. Data in Oahu were converted from polygon coverages to 30-meter grids. All GIS data in Central Columbia Plateau and Snake River Plain were converted to 1-kilometer grids because the percent land cover grids (Naomi Nakagaki, electronic commun.) were already mapped at 1-kilometer. Then, the logistic regression models similar to equation 1 were entered into a GIS and a probability rating was calculated for each of the grid nodes in each of the three study units.

Areas with a high percentage of land in crops (such as potatoes or sugarcane), a low percentage of fallow land, and highly permeable soils with low amounts of organic matter are most likely to have atrazine detected in the groundwater (fig. 33). Areas where agricultural activities were absent had much lower probabilities of atrazine being detected. The Snake River Plain had a much higher probability of atrazine detections, with more than 50 percent of the land area having greater than a 50-percent probability of atrazine contamination. Oahu had a much lower probability of atrazine contamination, with only 24 percent of the land area having greater than a 50-percent probability of atrazine contamination.

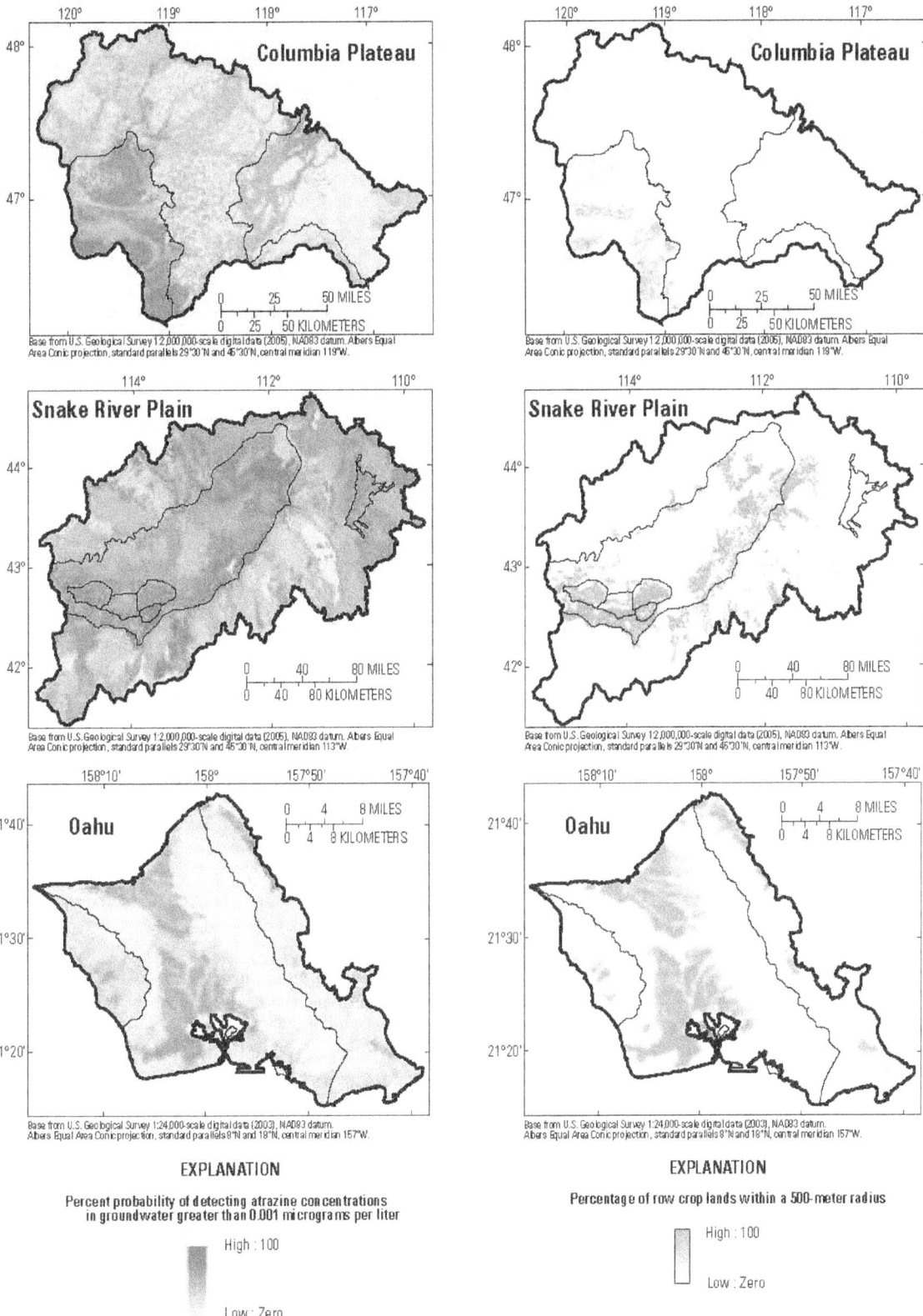

EXPLANATION

Percent probability of detecting atrazine concentrations
in groundwater greater than 0.001 micrograms per liter

High : 100

Low : Zero

EXPLANATION

Percentage of row crop lands within a 500-meter radius

High : 100

Low : Zero

Figure 33.   Probability from the model that atrazine will be detected in groundwater of the Columbia Plateau, Snake River Plain, and Oahu and percentage of row crops grown within a 500-meter radius in those areas.

Atrazine and deethylatrazine concentrations and frequencies of detections tended to be greater in younger groundwater (fig. 34). This is because atrazine was first registered for use in the United States by the USEPA in 1958, and it was not widely used in the United States until the mid-1960s (U.S. Environmental Protection Agency, 2003d). Atrazine use on crops increased over time, so its occurrence and concentrations in groundwater increased. The youngest groundwater is commonly located in the uppermost portions of an aquifer. It can take many years for groundwater contamination to move into deeper parts of an aquifer. Relations such as those shown in figure 34 are important, because they indicate that it can take many years for contamination to travel into the aquifer, and that contaminants can persist in groundwater for many decades.

The year the groundwater was recharged was determined in water samples from a subset of wells sampled in the Snake River Plain and Oahu using chlorofluorocarbons, tritium, and sulfur hexafluoride (Plummer and others, 2000; Hunt, 2004). Estimates of the year the groundwater was recharged have some error associated with them because of mixing of old waters and young waters within the well bore during sampling (Plummer and others, 2006). This mixing of old and young groundwater produces a mean groundwater age of the sample. The atrazine/deethylatrazine detections in waters older than 1960 in figure 34 are probably because of this mixing of older and younger groundwater; the atrazine or deethylatrazine probably is present in the young fraction of groundwater, even though the mean recharge date of the water as a whole is older than 1960.

Volatile Organic Compounds

Fourteen volatile organic compounds (VOCs) were detected in 5 percent or more of water samples from any one of the three study areas (fig. 35); 29 VOCs were detected altogether. The most frequently detected VOCs were fumigants and solvents; chloroform may originate as a chlorination byproduct. Oahu had the highest percentage of groundwater samples with VOC detections, both at any concentration (blue bars, including low "trace" concentrations) and at or above a common assessment level of 0.2 micrograms per liter (red bars). Detection rates at any concentration were high for Oahu in part because of lower reporting levels (see following section, "VOC Reporting Levels").

At the common assessment level of 0.2 micrograms per liter (red bars on figure 35), VOCs were far more prevalent in Oahu groundwater samples than in samples from the Columbia Plateau and Snake River Plain. VOC detections on Oahu were as high as 36 percent for the fumigant compound 1,2,3-trichloropropane (TCP) and 24 percent for the solvent trichloroethene (TCE). Other VOCs detected in 5 percent

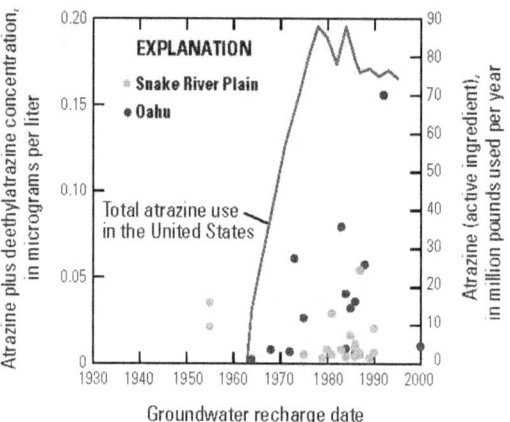

Figure 34. Atrazine and deethylatrazine concentrations versus groundwater recharge date for the Snake River Plain and Oahu and total atrazine use in the United States.

or more of samples include the solvent and disinfection byproduct chloroform (also known as trichloromethane), solvents perchloroethene (PCE) and carbon tetrachloride (tetrachloromethane), and the fumigant 1,2-dichloropropane (DCP), which was the most frequently detected VOC in the Columbia Plateau and Snake River Plain at and above the common assessment level.

Fumigant contamination in all three study areas is thought to have originated as nonpoint pollution from widespread agricultural application. Solvent contamination at high concentrations on Oahu likely originated from military sources, whereas origins for the fewer solvent detections in the Columbia Plateau and Snake River Plain are less clear and could range from urban sources to use on farm machinery to impurities in fumigant or pesticide formulations.

Chloroform was the most frequently detected VOC, present in 56 percent of Oahu samples. Chloroform is a compound in the trihalomethane class of VOCs (also known as disinfection byproducts). Its presence in groundwater is usually attributed to infiltration of chlorinated water from lawn or landscape irrigation. On Oahu, however, highest concentrations of chloroform were in water samples that also had high concentrations of TCE and PCE solvents, whereas low chloroform concentrations were codetected with moderate to low concentrations of fumigants and pesticides in agricultural areas where chlorinated water would not be used for irrigation. Judging from Oahu results, chloroform is present in solvent formulations and likely is present in fumigant or pesticide formulations as an additive or manufacturing byproduct or impurity (Hunt, 2004). This may explain many of the chloroform detections in the Columbia Plateau and Snake River Plain as well.

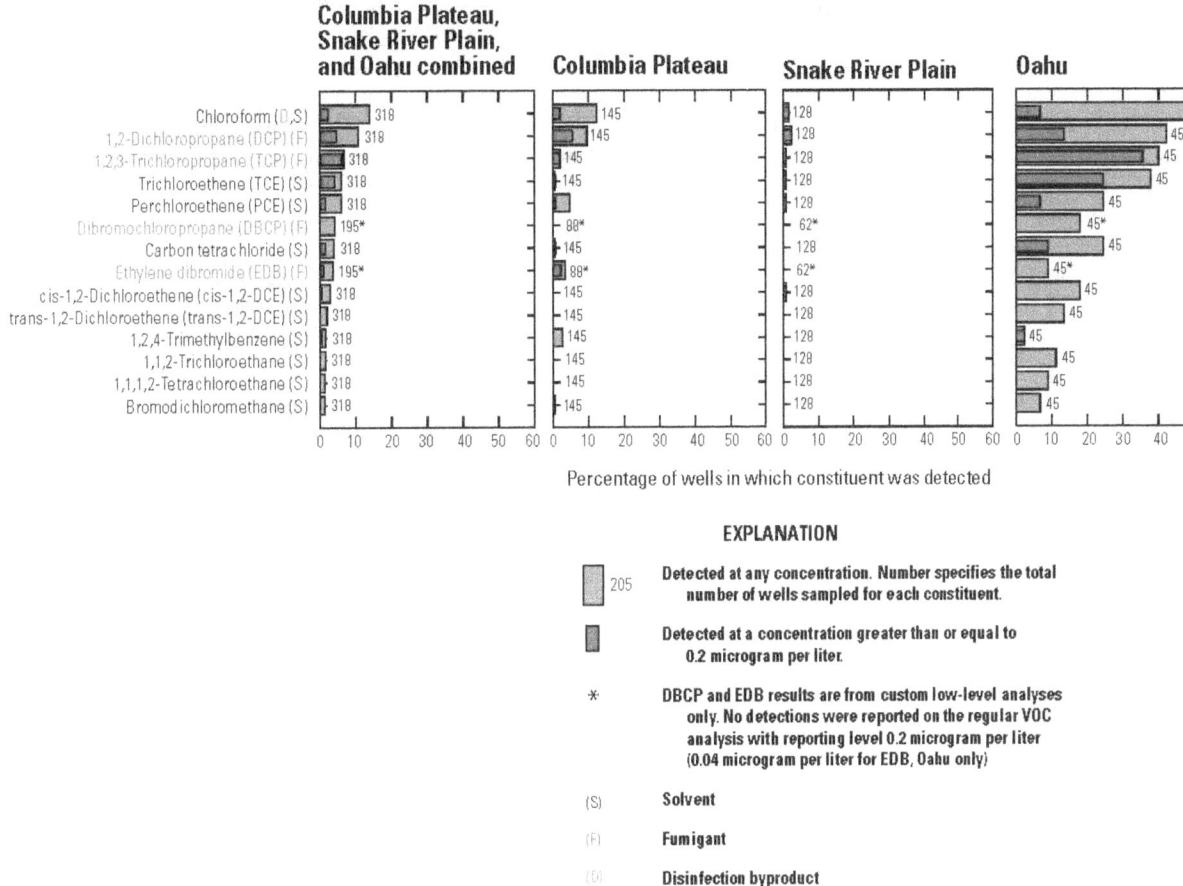

Figure 35. Percentage of groundwater samples with detections of selected volatile organic compounds, at any concentration (includes very low "trace" concentrations) and at or above a common assessment level of 0.2 micrograms per liter.

VOC Reporting Levels

A common assessment level is necessary for valid comparison among the three study areas because laboratory reporting levels and practices have varied during the two decades of NAWQA data collection. In the 1992–95 time span of NAWQA Cycle 1 sampling in the Columbia Plateau and Snake River Plain, the VOC laboratory analysis had a fixed minimum reporting level of 0.2 µg/L for all compounds analyzed. "Fixed" means that no concentrations were reported below the 0.2 µg/L reporting limit (values were reported as "less than 0.2"). By the time of Oahu's sampling in 2000–2001, refinements in the laboratory method had lowered reporting levels nearly an order of magnitude, to as low as 0.03 µg/L for some compounds. Furthermore, concentrations below reporting limits were reported as estimated concentrations, as low as 0.01 µg/L (some laboratories would report these as "trace" levels). The low reporting levels

revealed widespread VOC contamination on Oahu, but mostly at low concentrations below human-health benchmarks for drinking water. Similar low-level contamination may be present in the Columbia Plateau and Snake River Plain and may be discovered as well networks are resampled with the improved VOC laboratory method. Cycle 2 resampling of two Columbia well networks (a major-aquifer network ccptsus1b, and an orchard monitor-well network, ccptlusor1b) contributed a small subset of low-level VOC results (note the blue bars in the Columbia graph of figure 35). However, the only VOC results from the Snake River Plain were at the higher fixed reporting level of 0.2 µg/L, and it is this concentration that was chosen as a common assessment level for comparing across all three study areas. Oahu VOC results parallel those nationally, where VOCs were detected in 20 percent of wells at 0.2 µg/L but in more than 50 percent of wells at an assessment level an order of magnitude lower, at 0.02 µg/L (Zogorski and others, 2006).

VOC Detection Rates by Network and Well Type

Among drinking-water wells (fig. 36A), VOCs were detected in 5 percent or fewer of domestic wells but in as many as 30 percent of public-supply wells. Domestic wells were sampled only in the Columbia Plateau and Snake River Plain but not on Oahu. Many public-supply wells were sampled in the Columbia Plateau and Oahu, but only a few were sampled in the Snake River Plain. So the domestic-well graph in figure 36A has no Oahu representation and the public-supply well graph has little Snake River Plain representation and is highly skewed by results from Oahu, where VOCs were much more prevalent in groundwater than in the Columbia Plateau and Snake River Plain (see fig. 35). Domestic wells likely have lower VOC detection rates because they draw water from a small area of aquifer and many of the wells are on rural farm homesteads with few conspicuous sources of solvents and other VOC compounds (with the notable exception of fumigants). Public-supply wells with large cones of depression likely have higher VOC detection rates because they draw water from greater distances in the aquifer, thereby having a greater chance of intercepting contamination from solvent leaks or agriculturally applied fumigants. This is particularly true of Oahu wells, which make up 23 of the public-supply wells in figure 36A and are known to be affected by solvents used on military bases and agricultural fumigants.

Similarly, VOCs were detected at much higher rates in major-aquifer studies than in agricultural land-use studies (fig. 36B). But again, results are closely related to well type. The agricultural land-use graph only includes wells from the Columbia Plateau and Snake River Plain (mostly domestic and monitoring wells) because no land-use studies were conducted on Oahu. Moreover, the major-aquifer graph only includes wells from the Columbia Plateau and Oahu (mostly public-supply wells) because VOCs were not analyzed in Snake River Plain major-aquifer studies, only in the Snake River Plain land-use studies. The major-aquifer graph therefore reflects samples from public-supply wells, and high rates of detection can again be attributed to the wider capture of groundwater by the high-capacity wells and by the outsize influence of Oahu wells, previously shown to have high rates of contamination by solvents used on military bases and agricultural fumigants.

Sources of VOCs

Volatile organic compounds form a broad class of organic compounds that include solvents and degreasers, refrigerants, fumigants, gasoline components, and trihalomethanes, which can form when water is chlorinated for disinfection. Many of the compounds have multiple uses; for example, some solvent-class compounds can also originate as manufacturing byproducts, impurities in other solvents or as additives in pesticide formulations.

Groundwater can be contaminated by VOCs from a number of sources: accidental spills, improper disposal, chronic leakage from point sources such as storage tanks, and nonpoint application of chemical products at the land surface such as for agriculture. VOCs can even dissolve into water from the atmosphere, where they are present at generally low concentrations from industrial and manufacturing releases.

The most prominent use of VOCs common to all three study areas is nonpoint agricultural application of soil fumigants to combat nematodes (root worms). This is discussed in more detail in the following "Fumigants" section of this report. In addition to nonpoint application, point sources of fumigants have been identified at sites where fumigants were spilled during mixing and preparation.

Another prominent use of VOCs specific to Oahu is military use of solvents at bases in the central Oahu plateau. Solvent use for aircraft and automotive degreasing dates back to the World War II era of the 1940s. A greater awareness and environmental concern about chemical releases evolved over subsequent decades. Solvent disposal at automotive and machine shops often was to unlined earth or to drains that routed to sumps and cesspools. Landfills also have been documented sources of disposed solvents, typically in drums (Harding Lawson Associates, 1995; U.S. Environmental Protection Agency, 2000).

VOC use in the urban environment has included specific industries such as dry cleaning and manufacturing, as well as various incidental uses in household products by the general population. Gasoline is a source of several VOCs known as BTEX compounds (benzene, toluene, ethylbenzene, and xylene). Point releases of gasoline to the environment have been from spills and from chronic leakage from storage tanks and fuel-transmission pipelines.

Fumigants

Widespread agricultural use of soil fumigants, which are VOCs used as pesticides, in the Columbia Plateau, Snake River Plain, and Oahu has led to their detection in groundwater. Fumigant concentrations exceeded drinking-water benchmarks in all three study areas and caused well closures on Oahu in the 1980s.

NAWQA sampling detected fumigant compounds in only a few areas across the United States, corresponding to areas of known fumigant use. The Columbia Plateau, Snake River Plain, and Oahu are three such areas where fumigants have been applied extensively for agriculture. Although several fumigant compounds were banned from use during the late 1970s, they are still being detected in groundwater decades later.

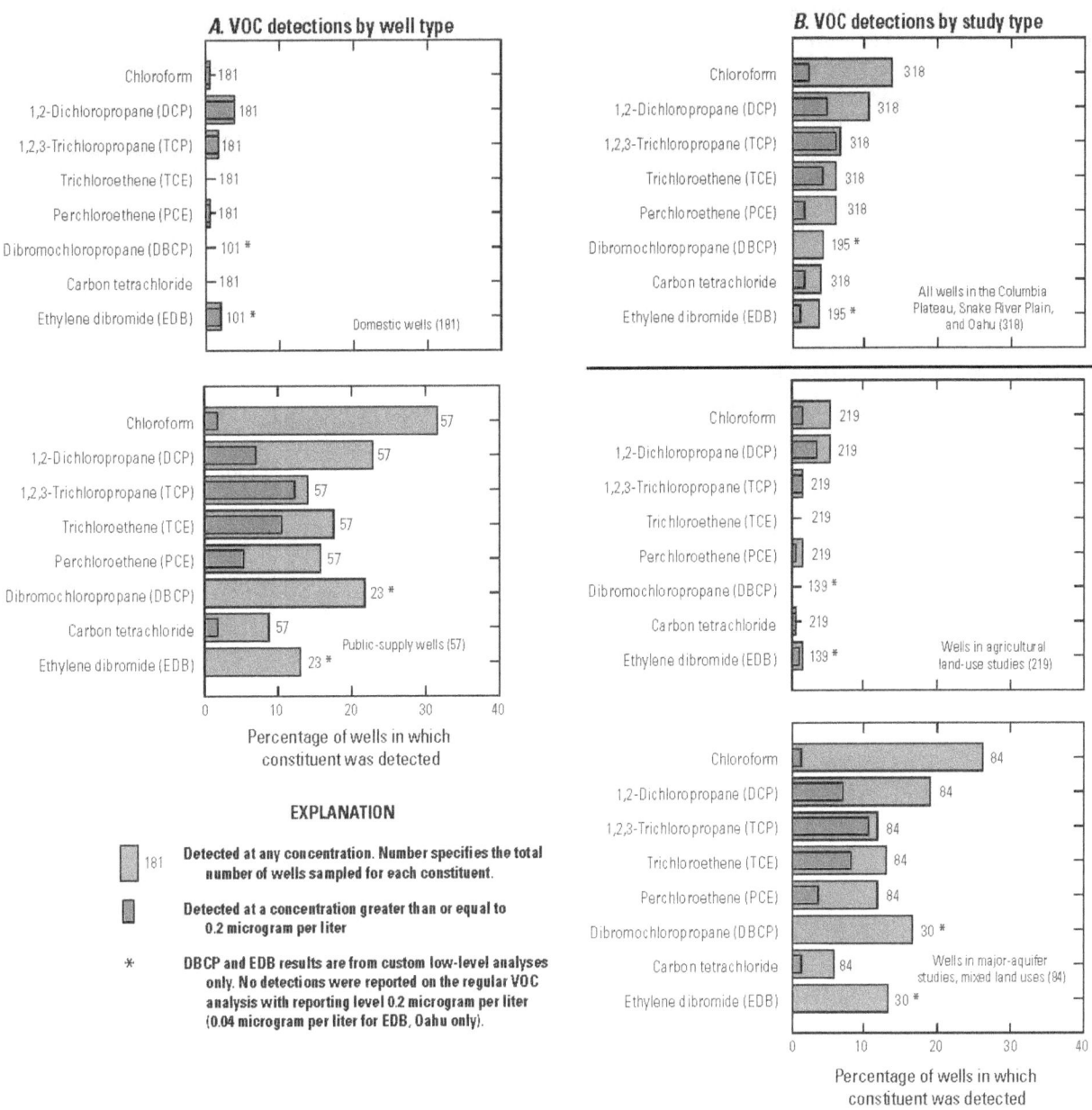

Figure 36. Percentage of wells with VOC detections for (A) public-supply and domestic drinking water wells and (B) wells in the land-use and major-aquifer study networks.

Fumigants were detected frequently in NAWQA sampling on Oahu, less frequently in the Columbia Plateau, and in water from only one well in the Snake River Plain (fig. 37 and table 4). Fumigants were detected in groundwater from 20 of the 45 wells sampled on Oahu: in 12 of 30 public-supply wells and in 8 of 15 monitoring wells (Hunt, 2004). Concentrations in some water samples were greater than Hawaii MCLs (Maximum Contaminant Levels; State of Hawaii, 2011), which are human-health benchmarks used to define and regulate safe drinking water. Of 30 samples from Oahu public-supply wells, 4 (13 percent) had fumigant concentrations greater than Hawaii MCLs for Dibromochloropropane (DBCP) or 1,2,3-Trichloropropane (TCP). Hawaii MCLs for these compounds are lower than USEPA MCLs.

Although fumigants were detected less frequently in the Columbia Plateau and Snake River Plain than on Oahu, they are still a concern in those areas because some concentrations were greater than human-health benchmarks. In the Columbia Plateau, two water samples had concentrations greater than the USEPA MCL for Ethylene dibromide (EDB) and the Columbia Plateau and Snake River Plain each yielded one water sample in which 1,2-Dichloropropane (DCP) concentration was greater than the USEPA MCL (table 4).

Fumigant Use

Soil fumigants are volatile organic compounds (VOCs) used as pesticides. They are applied to soils to reduce populations of plant parasitic nematodes (harmful rootworms), weeds, fungal pathogens, and other soil-borne microorganisms (U.S. Environmental Protection Agency, 2005, 2008). Being volatile, fumigants evaporate easily from liquid to vapor or gas (their other main agricultural use besides soil application is grain fumigation, where fumigant vapor is infused through stored grain to kill pests). Fumigants are injected or incorporated into the soil before planting crops, at which time the fumigant vapor permeates the soil and kills soil-borne pests. Much of the fumigant would evaporate to the atmosphere before having the desired effect, so the soil is compacted or plastic sheeting laid down to retain the chemical long enough for it to act. Soil fumigants are used across the

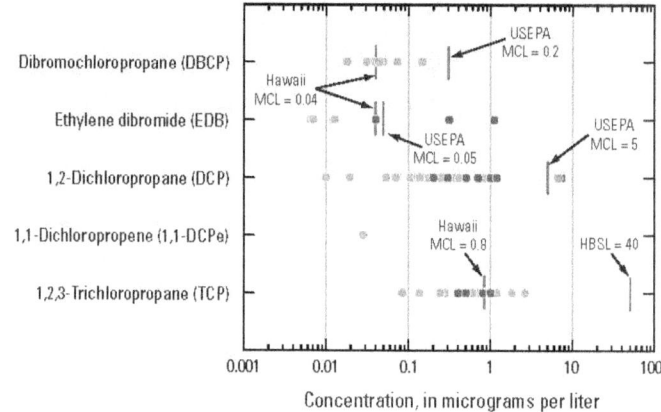

Figure 37. Fumigant concentrations detected in groundwater sampled from the Columbia Plateau, the Snake River Plain, and Oahu.

Nation on a variety of different crops (U.S. Environmental Protection Agency, 2005), but on Oahu they are used most commonly for pineapple, and in the Columbia Plateau and the Snake River Plain they are used most commonly for potatoes and sugar beets.

Fumigant use dates back to the 1940s and 1950s, when the compounds were first formulated and marketed commercially. Most of the fumigants detected in our three study areas were banned for soil application or usage was ceased in the late 1970s and early 1980s, and only 1,3-Dichloropropene is still used as a soil fumigant in the United States. DBCP was banned by the USEPA in 1979 after it was found to cause infertility in male workers exposed to it, and because of its potential to cause tumors in the breast, lung, and other organs in laboratory animals (Clark and Snedeker, 2004). EDB was banned in 1983 because it was found to be contaminating groundwater supplies in a number of States, and because laboratory test results had shown EDB to be a carcinogen and mutagen and to cause reproductive disorders in test animals (U.S. Environmental Protection Agency, 1983). DCP was withdrawn from soil fumigant formulations by manufacturers in the late 1970s and early 1980s. TCP, an impurity in DCP products, has been found to cause cancer in rodents (California Environmental Protection Agency, 2009).

Table 4. Fumigant detections in wells in the Columbia Plateau, Snake River Plain, and Oahu and their relation to human-health criteria.

[**Abbreviations:** µg/L, micrograms per liter (parts per billion); nd, none detected; USEPA, U.S. Environmental Protection Agency; MCL, Maximum Contaminant Level; HBSL, Health Based Screening Level; HMCL, Hawaii MCL; MCLs and HBSLs are human-health benchmarks for drinking water (for regulated and unregulated compounds, respectively)]

Fumigant compound	Columbia Plateau			Snake River Plain			Oahu			
	Number of wells sampled	Number of detections	Number of detections above USEPA human-health benchmarks	Number of wells sampled	Number of detections	Number of detections above USEPA human-health benchmarks	Number of wells sampled	Number of detections	Number of detections above USEPA human-health benchmarks	Number of detections above Hawaii human-health benchmarks
Dibromochloropropane (DBCP)	162	nd	nd	124	nd	nd	45	8	0 (MCL = 0.2 µg/L)	4 (HMCL = 0.04 µg/L)
Ethylene dibromide (EDB)	162	3	2 (MCL = 0.05 µg/L)	124	nd	nd	45	4	0 (MCL = 0.05 µg/L)	0 (HMCL = 0.04 µg/L)
1,2-Dichloropropane (DCP)	151	9	1 (MCL = 5 µg/L)	128	1	1 (MCL = 5 µg/L)	45	19	0 (MCL of 5 µg/L)	No human-health benchmark established
1,1-Dichloropropene (1,1-DCPe)[1]	151	nd	nd	128	nd	nd	45	1	No human-health benchmark established	No human-health benchmark established
1,2,3-Trichloropropane (TCP)	151	4	0 (HBSL = 40 µg/L)	128	nd	nd	45	18	0 (HBSL = 40 µg/L)	5 (HMCL = 0.8 µg/L)

[1] The detected compound, 1,1-Dichloropropene, is a breakdown product of 1,3-Dichloropropene (DCPe) used in fumigant formulations.

Millions of pounds of fumigants are used to produce crops every year in the United States (U.S. Environmental Protection Agency, 2005), although other chemicals have replaced the banned compounds. Methyl bromide, DCPE, metam sodium, and chloropicrin are the most widely used soil fumigants, and they rank in the top 20 pesticides (in pounds per year) because of high application rates and widespread coverage. In 2001, metam sodium was the third most commonly used pesticide in the United States (57 to 62 million pounds) and methyl bromide was the seventh most commonly used pesticide (20 to 25 million pounds). DCPE was the eighth most commonly used pesticide (20 to 25 million pounds), and chloropicrin was the eighteenth most commonly used pesticide (5 to 9 million pounds) (U.S. Environmental Protection Agency, 2004). Nationally, the largest uses of soil fumigants are on potatoes, tomatoes, tobacco, carrots, and strawberries.

Several of the fumigants have been determined to be highly carcinogenic, and this is reflected in the maximum contaminant levels for safe drinking water that have been set by USEPA and various States. USEPA MCLs for EDB (0.05 µg/L) and DBCP (0.2 µg/L) are the second and third lowest MCLs that have been established; only the MCL for dioxin is lower. As a result, even minute fractions of soil-applied fumigants can cause unsafe concentrations in groundwater. Detection of fumigants has resulted in closures of high-capacity public-supply wells and has required installation of costly carbon filtration to render the water drinkable.

Even though most of the fumigants investigated by this report have not been used since the early 1980s, they will continue to be detected in groundwater for many years into the future. Computer modeling by Rungvetvuthivitaya and others (2007) indicate that it can take from 14 to 32 years for DBCP and EDB to travel through the soil and bedrock to reach the groundwater in Oahu. After they enter the groundwater, they can persist for many years. For example, Burlinson and others (1982) predict the half-life time for DBCP in Oahu (the time that it takes for the concentration of the constituents to break down to one-half of the original concentration) to be 38 years if the groundwater temperature is 25 degrees Celsius and the pH is 7. If the groundwater temperature is 15 degrees Celsius, DBCP's half-life is 141 years (the average temperature of the groundwater sampled in Oahu for this Circular was 22 degrees Celsius and the average pH was 7.1). For these reasons, fumigants may be thought of as "legacy contaminants"— originating from surface application decades earlier rather than corresponding to current land use and chemical application practices.

Trace Elements

Trace elements include metals and semi-metallic elements that typically are found in natural waters at concentrations less than 1 mg/L. These elements originate primarily from rock weathering; concentrations of trace elements in groundwater reflect their abundance in aquifer materials, geochemical conditions, concentrations of other constituents, and attenuation processes such as adsorption. Many trace elements may occur as multiple ionic species in natural waters that, depending on redox conditions and pH, have different solubility characteristics. Human activities such as mining and waste disposal also can affect concentrations of trace elements in groundwater. At high concentrations, many trace elements can have adverse health effects, whereas others may present aesthetic or nuisance problems.

Trace metals that can have adverse health effects include lead, arsenic, and molybdenum. Long-term exposure to lead by infants and children can cause delays in physical or mental development; children could show slight deficits in attention span and learning abilities. Arsenic, which has been recognized as a toxic element for centuries, is a human health concern because it can contribute to skin, bladder, and other cancers (National Research Council, 1999). Arsenic is a naturally occurring element in rocks and soils, and therefore in the groundwater that is in contact with them (Welch and others, 2000). Elevated arsenic concentrations in groundwater can occur naturally, from mobilization of arsenic into groundwater as a result of irrigation (Ayotte and others, 2011) or from the use of arsenic-containing pesticides. Molybdenum is considered an essential trace element in both animals and humans (World Health Organization, 2011). However, consequences of long-term consumption of drinking water with high concentrations of molybdenum can include enlarged liver, disorders of the gastrointestinal tract, kidneys, and a gout-like disease (joint pain in the hands and feet) (Wisconsin Department of Health Services, 2010).

Trace elements were not analyzed in samples from all of the wells sampled in this study. The concentrations of nine trace elements—arsenic, cadmium, chromium, copper, lead, selenium, zinc, iron, and manganese—were analyzed in about 260 wells. The remaining trace elements were analyzed in about 180 of the sampled wells.

Many trace elements were minor constituents in the well water sampled in this study. Aluminum, arsenic, barium, boron, chromium, copper, lithium, molybdenum, nickel, strontium, uranium, vanadium, and zinc were each detected in about one-half or more of the sampled wells (using a common reporting level of 1 µg/L for most trace elements, fig. 38). Strontium, barium, boron, and vanadium were the most frequently detected and occurred in almost all samples. Arsenic, uranium, and molybdenum, which have health concerns associated with them, were all detected in more than 60 percent of samples. Antimony, beryllium, cobalt, cadmium, silver, and thallium were not detected (using a common reporting level of 1 µg/L).

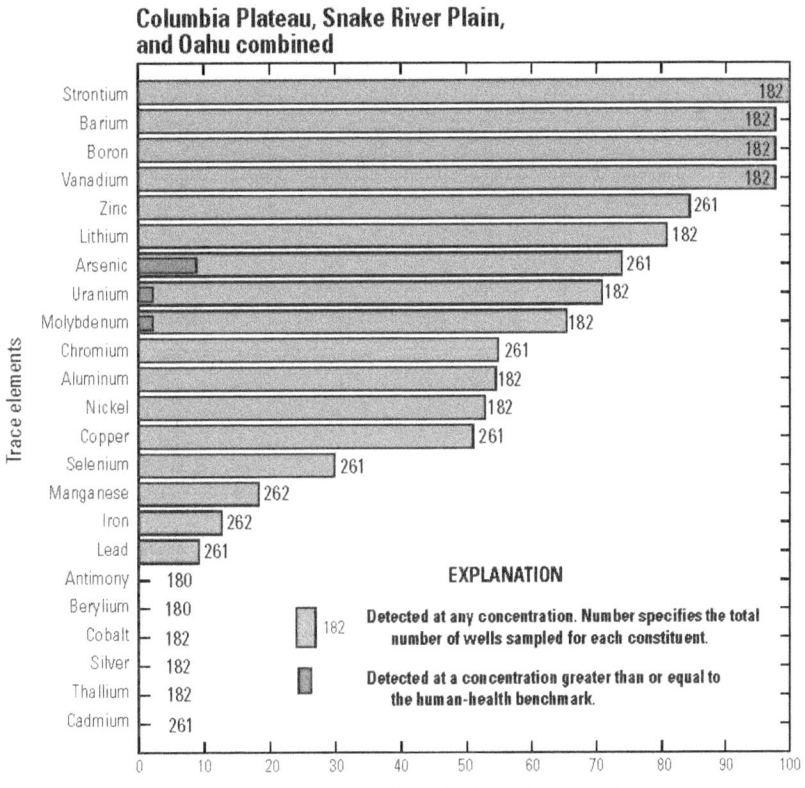

Figure 38. Percentage of wells in the Columbia Plateau, Snake River Plain, and Oahu with trace-element detections.

Radon

Radon (radon-222) is a water-soluble, radioactive gas that originates from radium-226, part of the uranium-238 decay series. Because its initial parent uranium occurs ubiquitously in trace amounts in the aquifer sediments and rocks, and because it is soluble, radon is common in groundwater. Radon is chemically inert and does not react with aquifer materials or other chemical constituents, but it decays through alpha-particle emission and has a short half-life (3.8 days; Wanty and Nordstrom, 1993). Radon concentrations in groundwater can be affected by multiple factors, including the distribution of uranium-bearing minerals in aquifer materials, aquifer physical characteristics, and geochemical conditions that affect the uranium and radium mobility (Hess and others, 1985; Otton, 1992; Wanty and others, 1992). Radon concentrations are reported in activity units, picocuries per liter (pCi/L), which describe the number of radioactive emissions (nuclear disintegrations) over time, rather than in mass concentration units. One picocurie per liter equals 2.2 radioactive disintegrations per minute per liter.

Radon and other naturally occurring radionuclides emit ionizing radiation and consequently are carcinogens. Radon can contribute to the risk of developing lung and gastrointestinal cancers (National Academy of Sciences, 1999). According to the U.S. Environmental Protection Agency (2009b), radon is the second most frequent cause of lung cancer, after cigarette smoking, causing 21,000 lung cancer deaths per year in the United States. Research has shown that the potential of lung cancer from breathing radon in air is much larger than the potential of stomach cancer from swallowing drinking water with radon in it (U.S. Environmental Protection Agency, 2009b). Most of the exposure from radon in drinking water comes from radon released into the air when water is used for showering and other household purposes. Adverse health effects from radon in drinking water result primarily from inhalation, after the gas is released from solution in the home, although the contribution from drinking water usually is small compared to other sources of radon in indoor air (Hopke and others, 2000). Water with about 10,000 pCi/L of radon contributes about 1 pCi/L of radon to indoor air (Otton, 1992); USEPA recommends that homes with indoor air concentrations at or above 4 pCi/L be fixed to reduce concentrations (U.S. Environmental Protection Agency, 2009b). Two human-health benchmarks, which are regulations proposed by USEPA in 1999 for public water systems, are used for comparison with radon concentrations in this study. The higher value,

4,000 pCi/L, is an alternative MCL that is proposed for public water systems for states or water-system service areas that have programs in place to reduce radon risks from all sources (Hopke and others, 2000; U.S. Environmental Protection Agency, 2011d). The lower value, 300 pCi/L, is proposed as the MCL for states or service areas that do not have such programs.

Radon concentrations were measured in nearly all of the wells sampled in this study. Radon activities in water exceeded the proposed human-health benchmark of 4,000 pCi/L in very few of the wells sampled in the Columbia Plateau, Snake River Plain, and Oahu (fig. 39). However, using the lower USEPA proposed human-health benchmark of 300 pCi/L, radon activities exceeded the human-health benchmark in water from over 80 percent of the wells sampled in the Columbia Plateau and 50 percent of the wells sampled in the Snake River Plain. Less than 5 percent of wells sampled in Oahu exceeded the lower human-health benchmark of 300 pCi/L, indicating that there is a much lower potential of radon exposure from groundwater in Oahu.

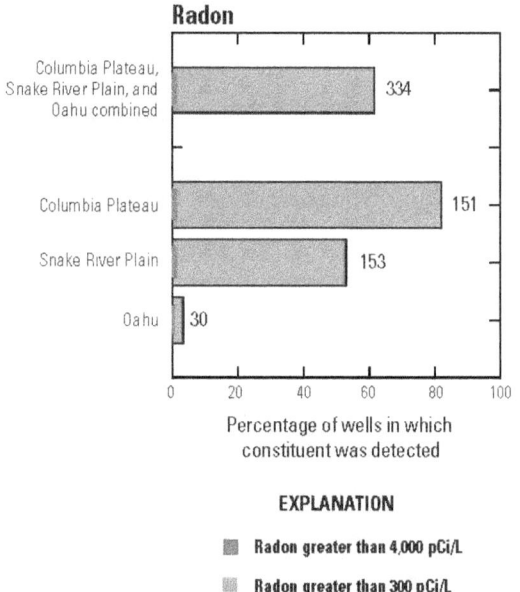

EXPLANATION

▨ Radon greater than 4,000 pCi/L

▨ Radon greater than 300 pCi/L

Figure 39. Percentage of wells with radon detections above human-health benchmarks in water from wells sampled in the Columbia Plateau, Snake River Plain, and Oahu.

Relation of Constituent Concentrations to Human-Health Benchmarks

Concentrations of naturally occurring and anthropogenic (related to human activities) constituents were detected above human-health benchmarks in drinking water from domestic and public-supply wells sampled in the Columbia Plateau, Snake River Plain, and Oahu (fig. 40 and table 5). Naturally occurring compounds detected above human-health benchmarks are radon, arsenic, and molybdenum. Anthropogenic compounds detected above human-health benchmarks are nitrate, dieldrin, Ethylene dibromide (EDB),

Dibromochloropropane (DBCP), 1,2-Dichloropropane (DCP), 1,2,3-Trichloropropane (TCP), and Trichloroethene (TCE). Naturally occurring constituents and nitrate concentrations above human-health benchmarks were more common in the Columbia Plateau and the Snake River Plain. Anthropogenic constituents above human-health benchmarks were more common in Oahu than the Columbia Plateau or Snake River Plain. The percentage of wells exceeding 10 percent of the human health benchmarks provides an indication of contaminants that may approach concentrations of potential human-health concern, either individually or as mixtures, and to identify those that may warrant additional monitoring and study (table 5).

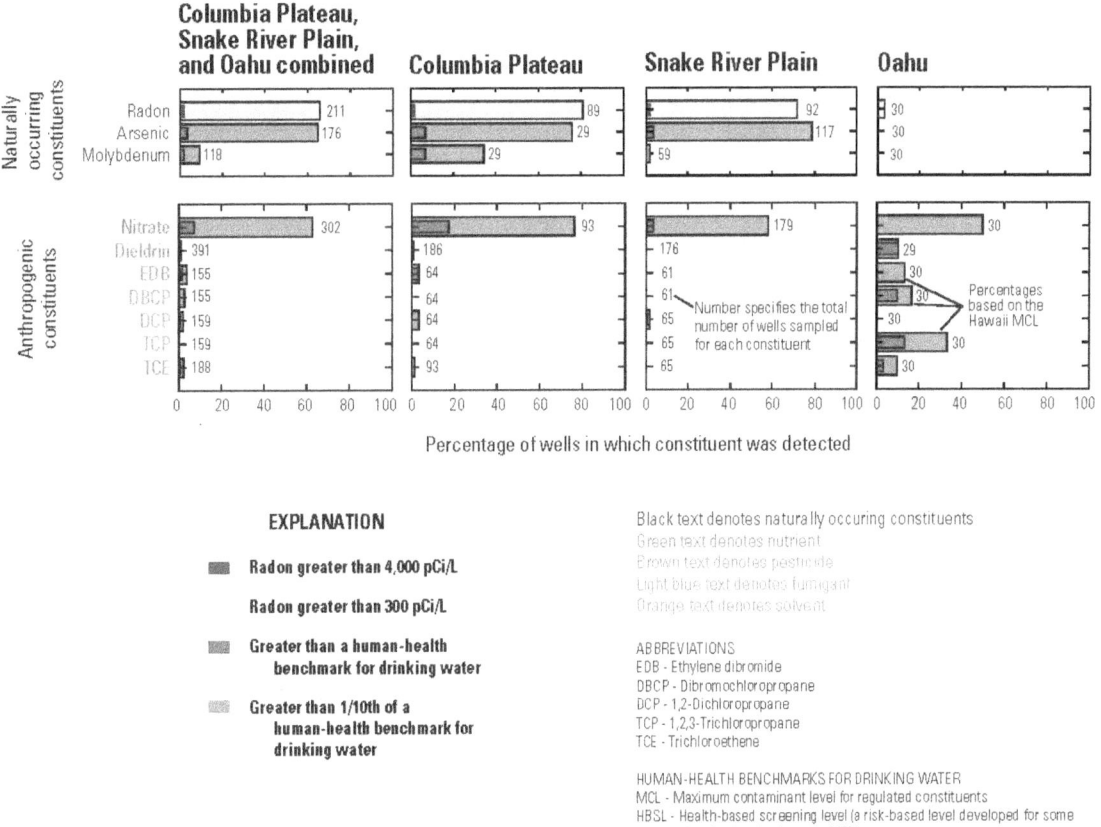

Figure 40. Percentage of wells with detections of selected constituents above human-health benchmarks in water from domestic and public-supply wells sampled in the Columbia Plateau, Snake River Plain, and Oahu.

Table 5.   Percentage of wells with water containing constituents exceeding human-health criteria in the Columbia Plateau, Snake River Plain, and Oahu.

[Abbreviations: Nitrate, nitrite plus nitrate as nitrogen; USEPA, U.S. Environmental Protection Agency; pCi/L, picocuries per liter; mg/L, milligrams per liter; µg/L, micrograms per liter; MCL, maximum contaminant level established by the USEPA; SMCL, secondary maximum contaminant level established by the USEPA; HBSL, health-based screening level developed by the USGS for constituents that have no MCL or SMCL; TT action level, USEPA treatment technique that requires public water systems to control the corrosiveness of their water]

Constituent	Human health benchmark[1] — Value	Human health benchmark[1] — Type	The Columbia Plateau, Snake River Plain, and Oahu combined — Number of wells sampled	Combined — Percent exceeding Human-health benchmark	Combined — Percent exceeding One tenth of human-health benchmark	Columbia Plateau — Number of wells sampled	Columbia Plateau — Percent exceeding Human-health benchmark	Columbia Plateau — Percent exceeding One tenth of human-health benchmark	Snake River Plain — Number of wells sampled	Snake River Plain — Percent exceeding Human-health benchmark	Snake River Plain — Percent exceeding One tenth of human-health benchmark	Oahu — Number of wells sampled	Oahu — Percent exceeding Human-health benchmark	Oahu — Percent exceeding One tenth of human-health benchmark
Naturally occurring constituents														
Radon	300 pCi/L	Proposed MCL[2]	211	66	100	89	81	100	92	72	100	30	3	100
	4,000 pCi/L	Proposed MCL[2]												
Arsenic	10 µg/L	MCL[2]	211	1	55	89	1	65	92	1	63	30	0	3
Fluoride	4 mg/L	MCL	176	3	65	29	7	76	117	3	79	30	0	0
Uranium	30 µg/L	MCL	300	0	46	92	0	47	178	0	54	30	0	0
Molybdenum	40 µg/L	HBSL	118	2	26	29	7	34	59	0	25	30	0	20
Lead	15 µg/L	TT Action Level[3]	176	0	9	29	0	3	117	0	2	30	0	0
Anthropogenic constituents — Nutrient														
Nitrate	10	MCL	302	7	63	93	17	76	179	3	58	30	0	50
Pesticide														
Atrazine	3 µg/L	MCL	392	0	2	186	0	2	177	0	2	29	0	0
Dieldrin	0.002 µg/L	HBSL low[4]	391	1	1	186	1	1	176	0	0	29	10	10
Bromacil	70 µg/L	HBSL	269	0	0	64	0	0	176	0	1	29	0	0
Fumigant														
Ethylene dibromide (EDB)	0.05 µg/L	MCL	155	0	4	64	0	3	61	0	0	30	[5]0	[5]13
Dibromochloropropane (DBCP)	0.2 µg/L	MCL	155	3	4	64	3	0	61	0	0	30	[6]10	[6]17
1,2-Dichloropropane (DCP)	5 µg/L	MCL	159	1	2	64	0	3	65	2	2	30	0	0
1,2,3-Trichloropropane (TCP)	40 µg/L	HBSL	159	2	6	64	0	0	65	0	0	30	[7]13	[7]33
Solvent														
Trichloroethene (TCE)	5 µg/L	MCL	188	0	2	93	0	1	65	0	0	30	3	10
Perchloroethene (PCE)	5 µg/L	MCL	188	0	2	93	0	1	65	0	2	30	0	3
Tetrachloromethane (carbon tetrachloride)	5 µg/L	MCL	188	0	1	93	0	1	65	0	0	30	0	3

Table 5. Percentage of wells with water containing constituents exceeding human-health criteria in the Columbia Plateau, Snake River Plain, and Oahu.—Continued

[**Abbreviations:** Nitrate, nitrite plus nitrate as nitrogen; USEPA, U.S. Environmental Protection Agency; pCi/L, picocuries per liter; mg/L, milligrams per liter; μg/L, micrograms per liter; MCL, maximum contaminat level established by the USEPA; SMCL, secondary maximum contaminant level established by the USEPA; HBSL, health-based screening level developed by the USGS for constituents that have no MCL or SMCL; TT action level, USEPA treatment technique that requires public water systems to control the corrosiveness of their water]

[1] Human health benchmarks are concentrations of constituents in drinking water that may be of potential concern for human health, if exceeded. The USEPA has established MCLs and SMCLs for some constituents (U.S. Environmental Protection Agency, 2011a). The USGS has established HBSLs for additional constituents that do not have MCLs or SMCLs established (Toccalino and others, 2008).

[2] There is currently no Federally-enforced drinking water standard for radon. EPA has proposed to require community water suppliers to provide water with radon levels no higher than 4,000 pCi/L, which contributes about 0.4 pCi/L of radon to the air in your home from showering and other household uses. Under the proposed regulation, States that choose not to develop enhanced indoor air programs will be required to reduce radon levels in drinking water to 300 pCi/L. This amount of radon in water contributes about 0.03 pCi/L of radon to the air in your home. (See: water.epa.gov/lawsregs/rulesregs/sdwa/radon/basicinformation.cfm and http://water.epa.gov/lawsregs/rulesregs/sdwa/radon/regulations.cfm).

[3] Lead is regulated by a USEPA Treatment Technique that requires systems to control the corrosiveness of their water. If more than 10 percent of tap water samples exceed 0.015 mg/L, water systems must take additional steps.

[4] There are two HBSLs established for dieldrin, 0.002 μg/L and 0.2 μg/L. The low value (0.002 μg/L) corresponds to a one in one million cancer risk, the high value (0.2) corresponds to a one in ten thousand cancer risk.

[5] The percent exceeding the human health benchmark for Oahu were computed using the Hawaii MCL of 0.04 μg/L. The percent exceeding the human health benchmark and one tenth of the human health benchmark for the Columbia Plateau, Snake River Plain, and Oahu combined, the Columbia Plateau, and the Snake River Plain were computed using the USEPA MCL of 0.05 μg/L.

[6] The percent exceeding the human health benchmark for Oahu were computed using the Hawaii MCL of 0.04 μg/L. The percent exceeding the human health benchmark and one tenth of the human health benchmark for the Columbia Plateau, Snake River Plain, and Oahu combined, the Columbia Plateau, and the Snake River Plain were computed using the USEPA MCL of 0.2 μg/L.

[7] The percent exceeding the human health benchmark for Oahu were computed using the Hawaii HBSL of 0.8 μg/L. The percent exceeding the human health benchmark and one tenth of the human health benchmark for the Columbia Plateau, Snake River Plain, and Oahu combined, the Columbia Plateau, and the Snake River Plain were computed using the HBSL of 40 μg/L.

Nitrate

Nitrate concentrations of water from drinking water wells were above the human-health benchmark of 10 mg/L in the Columbia Plateau (17 percent of wells) and the Snake River Plain (3 percent of wells) but did not exceed the human-health benchmark in Oahu (fig. 40). Nitrate can be both naturally occurring and from anthropogenic sources. Rupert (1996) reported that naturally occurring nitrate concentrations in the Snake River Plain are below 1 mg/L; Dubrovsky and others (2010) reported that national background concentrations of nitrate in groundwater are less than 1 mg/L; and Nolan and Hitt (2003) reported that nitrate concentrations of groundwater in relatively undeveloped areas of the United States are below 1.1 mg/L. The source of nitrate concentrations above background levels in groundwater of the Columbia Plateau is primarily nitrogen fertilizers, and to a lesser extent animal manure (Jones and Wagner, 1995). The sources of nitrate in groundwater of the Snake River Plain also are nitrogen fertilizers and animal manure (Rupert, 1996; Skinner and Donato, 2003). Domestic septic systems account for less than 1 percent of the total nitrogen input in the Snake River Plain (Rupert, 1996). Nitrate concentrations of water from public-supply wells sampled in Oahu were above the background concentration of 1 mg/L, but they were below human-health benchmarks. The elevated nitrate concentrations above background in Oahu are the results of decades of agricultural activities, particularly in sugar cane fields (Hunt, 2004).

Nitrate concentrations above the human-health benchmark are a concern because ingesting nitrate in drinking water by infants can cause low oxygen levels in their blood (called Methemoglobinemia, or Blue Baby Syndrome). Long-term exposure to nitrate at concentrations of 2 to 4 mg/L in community water supplies has possible links to bladder and ovarian cancer (Weyer and others, 2001) and to a type of cancer called non-Hodgkins lymphoma (Ward and others, 1996).

Pesticides and VOCs

Anthropogenic constituents were more commonly detected in water from drinking water wells in Oahu than the Columbia Plateau or Snake River Plain (fig. 40). Although Oahu had lower concentrations of radon, trace elements, and nitrate than the Columbia Plateau or the Snake River Plain, Oahu had higher concentrations and higher occurrences of dieldrin, EDB, DBCP, DCP, TCP, and TCE. Soil fumigants EDB, DBCP, DCP, and TCP are used in Oahu primarily on pineapple crops to help control soil-borne pests and are in the volatile organic compound (VOC) chemical class.

Dieldrin is an insecticide that was formerly used for termite control in Oahu. Dieldrin was applied as a parent compound and can also be a breakdown product of the termiticide aldrin, which was also used in Oahu (Brasher and Anthony, 2000). Dieldrin was detected in groundwater above the human-health benchmark of 0.002 µg/L from 10 percent of the public-supply wells sampled in Oahu (table 5). Originally developed in the 1940s as an alternative to DDT, dieldrin proved to be a highly effective insecticide and was very widely used during the 1950s to early 1970s. However, dieldrin is an organic pollutant that does not easily break down over time and tends to biomagnify (U.S. Environmental Protection Agency, 2003c). Long-term exposure has proven toxic to a very wide range of animals, including humans—far greater than to the original insect targets (U.S. Environmental Protection Agency, 2003c). For this reason, it is now banned in most of the world. Dieldrin has been linked to health problems such as Parkinson's disease, breast cancer, and immune, reproductive, and nervous system damage (U.S. Environmental Protection Agency, 2003c). It can also adversely affect testicular descent in the fetus if a pregnant woman is exposed to Dieldrin (U.S. Environmental Protection Agency, 2003c).

Trichloroethylene was detected above the human-health benchmark of 5 µg/L in water from less than 5 percent of public-supply wells in Oahu. Tetrachloroethylene and tetrachloromethane (carbon tetrachloride) were also detected in water from public-supply wells in Oahu, but below human-health benchmarks (fig. 40).

Radon

Radon activities in water from drinking water wells exceeded the proposed human-health benchmark of 4,000 pCi/L in only 1 percent of the wells sampled in the Columbia Plateau and Snake River Plain and in none of the wells sampled in Oahu (fig. 40). However, USEPA also has proposed a lower human-health benchmark of 300 pCi/L. Using this lower human-health benchmark, radon activities exceeded the proposed human-health benchmark in water from more than 80 percent of the wells sampled in the Columbia Plateau and the Snake River Plain. Only 3 percent of public-supply wells sampled in Oahu exceeded the lower human-health benchmark of 300 pCi/L.

Trace Metals

Uranium concentrations were below the human-health benchmark of 30 µg/L, but exceeded one-tenth of the human-health benchmark (3 µg/L), in water from 55 percent of the drinking water wells sampled in the Columbia Plateau and 25 percent of the wells sampled in the Snake River Plain (fig. 40). None of the water samples collected from public-supply wells in Oahu exceeded 1/10th of the human-health benchmark for uranium. Uranium occurrence was similar to radon occurrence in groundwater, which is to be expected because radon is derived from uranium. As indicated by uranium and radon concentrations in groundwater, uranium concentrations in the aquifer materials of Oahu are lower than concentrations in the Columbia Plateau and Snake River Plain.

Arsenic exceeded the human-health benchmark of 10 µg/L in water from 7 percent of the drinking water wells sampled in the Columbia Plateau and in water from 3 percent of the wells sampled in the Snake River Plain (fig. 40). However, arsenic concentrations exceeded 1/10th of the human-health benchmark in more than 75 percent of the drinking water wells sampled from the Columbia Plateau and Snake River Plain (table 5). Arsenic concentrations measured in water from public-supply wells in Oahu were lower than 1/10th of the human-health benchmark, so arsenic is not a concern in Oahu.

Lead concentrations in water from drinking water wells were below their human-health benchmark; it was rare to detect lead concentrations above 1/10th of the human-health benchmark (fig. 40). However, this study only collected fresh groundwater samples collected at the wellheads. In some cases, lead can leach from the pipes and other components of plumbing and distribution systems, increasing lead concentrations to above human-health benchmarks at the tap. For this reason, the USEPA specifies a treatment technique that requires public water systems to control the corrosiveness of their water. If more than 10 percent of tap water samples exceed lead concentrations of 0.015 mg/L, water systems must take additional steps.

Molybdenum exceeded the human-health benchmark of 40 µg/L in water from 7 percent of the drinking water wells sampled in the Columbia Plateau; water from 34 percent of those wells had molybdenum concentrations greater than 10 percent of the human-health benchmark (fig. 40). In sharp contrast, none of the water samples collected from drinking water wells in the Snake River Plain and Oahu exceeded the human-health benchmark, and only 2 percent of the water samples from the Snake River Plain exceeded 10 percent of the human-health benchmark.

Dissolved Solids, Sulfate, Chloride, Manganese, and Iron

Secondary maximum contaminant levels (SMCLs) were exceeded for dissolved solids, sulfate, chloride, manganese, and iron in water from drinking-water wells sampled in the Columbia Plateau, Snake River Plain, and Oahu (fig. 41 and table 6). Dissolved solids (also known as total dissolved solids, or TDS) is a measure of the combined content of all inorganic and organic constituents dissolved in the water sample. Dissolved solids are used as an indication of aesthetic characteristics of drinking water and as an aggregate indicator of the presence of a broad array of chemical constituents such as bromide, calcium, chloride, fluoride, iron, magnesium, manganese, nitrate, phosphorus, potassium, silica, sodium, and sulfate. Naturally occurring dissolved solids can arise from the weathering and dissolution of rocks and soils, but elevated dissolved solids can also result from agricultural and residential runoff, leaching of soil contamination and point source water pollution, discharge from industrial or sewage treatment plants, or runoff in snowy climates where road de-icing salts are applied. Dissolved solids concentrations tended to be higher in the Columbia Plateau and the Snake River than in Oahu (fig. 41).

Manganese and iron concentrations exceeded SMCLs much more commonly in the Columbia Plateau and the Snake River Plain than in Oahu. Manganese and iron can stain plumbing fixtures and can impart an unpleasant taste to the water, so they are undesirable constituents in drinking water. Elevated manganese and iron concentrations can occur in groundwater under reducing conditions, in groundwater with low dissolved oxygen, or in groundwater with low pH. Sulfate only exceeded the SMCLs in the Columbia Plateau; sulfate can impart a bad taste to the water and in high concentrations can cause dysentery problems. Chloride concentrations were a little higher in Oahu; chloride can impart an unpleasant taste and can cause corrosion problems in plumbing fixtures and delivery systems.

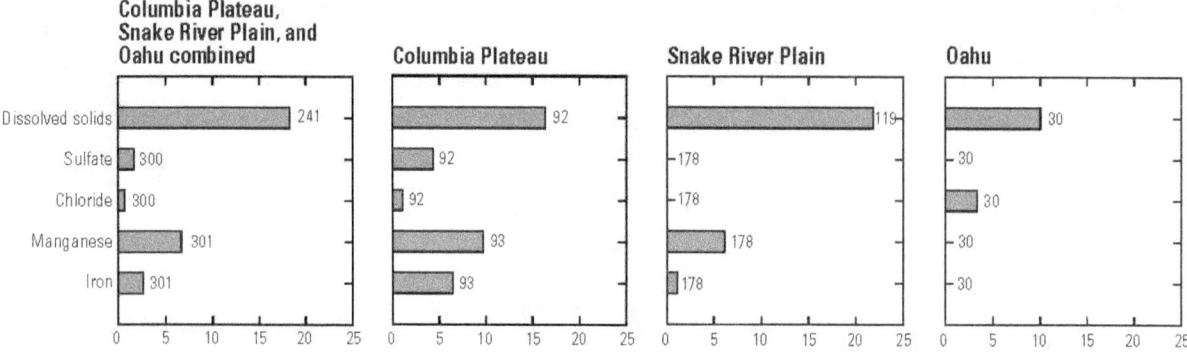

Percentage of wells in which constituent was detected above secondary maximum contaminant levels

EXPLANATION

☐ 30   Percentage of wells with concentrations greater than the
        Secondary Maximum Contaminant Level (SMCL). Number
        specifies the total number of wells sampled for each constituent

Figure 41.   Percentage of wells with concentrations greater than the secondary maximum contaminant levels (SMCL) for dissolved solids, sulfate, chloride, manganese, and iron in water from domestic and public-supply wells sampled in the Columbia Plateau, Snake River Plain, and Oahu.

Table 6.   Percentage of wells that exceeded secondary maximum contaminant levels for dissolved solids, sulfate, chloride, manganese, and iron in water from domestic and public supply wells sampled in the Columbia Plateau, Snake River Plain, and Oahu.

[**Abbreviations:** SMCL, secondary maximum contaminant level established by the U.S. Environmental Protection Agency; mg/L, milligrams per liter; µg/L micrograms per liter]

Constituent	SMCL value	The Columbia Plateau, Snake River Plain, and Oahu combined		Columbia Plateau		Snake River Plain		Oahu	
		Number of wells sampled	Percent exceeding SMCL	Number of wells sampled	Percent exceeding SMCL	Number of wells sampled	Percent exceeding SMCL	Number of wells sampled	Percent exceeding SMCL
Dissolved solids	500 mg/L	241	18	92	16	119	22	30	10
Sulfate	250 mg/L	300	2	92	4	178	0	30	0
Chloride	250 mg/L	300	1	92	1	178	0	30	3
Manganese	50 µg/L	301	7	93	10	178	6	30	0
Iron	300 µg/L	301	3	93	6	178	1	30	0

Natural Processes and Human Activities Affecting the Quality of Water in the Aquifers

Natural characteristics of the Columbia, Snake, and Oahu principal aquifers allow water and chemicals to infiltrate to the water table despite depths to water commonly in the hundreds of feet. The aquifers are essentially unconfined, having few extensive clay layers to impede infiltration through alluvial sediments and volcanic rock. Agriculture is intensive in all three study areas, and heavy irrigation has altered natural flow systems and fostered leaching of fertilizers and pesticides applied at land surface. The aquifers are mostly oxic, which hinders chemical breakdown of nitrate and many solvents but fosters breakdown of other constituents such as petroleum hydrocarbons.

This section summarizes the natural processes and human activities that affect groundwater as it recharges the aquifer and moves deeper in the system toward wells used for private, public supply, and irrigation.

Natural Aquifer Characteristics

Principal aquifers in Columbia Plateau, Snake River Plain, and Hawaii are highly vulnerable to contamination from the land surface. This results from natural properties of earth materials that make them susceptible to water and contaminant transport (well-drained soils, permeable volcanic rock, lack of regionally extensive clay layers) coupled with intensive agricultural and urban land use in which chemicals are applied or released at the land surface. Although some contaminants are natural products of rock weathering and plant decay, most constituents of concern in the study areas originate as land-applied chemicals or from chemical spills or improper disposal. Once released, contaminants join the flow of water along hydrologic pathways throughout the landscape (fig. 42). Surface runoff transports sediment and chemical residues to streams and rivers quickly. Infiltration to the deep water table may take a year or more.

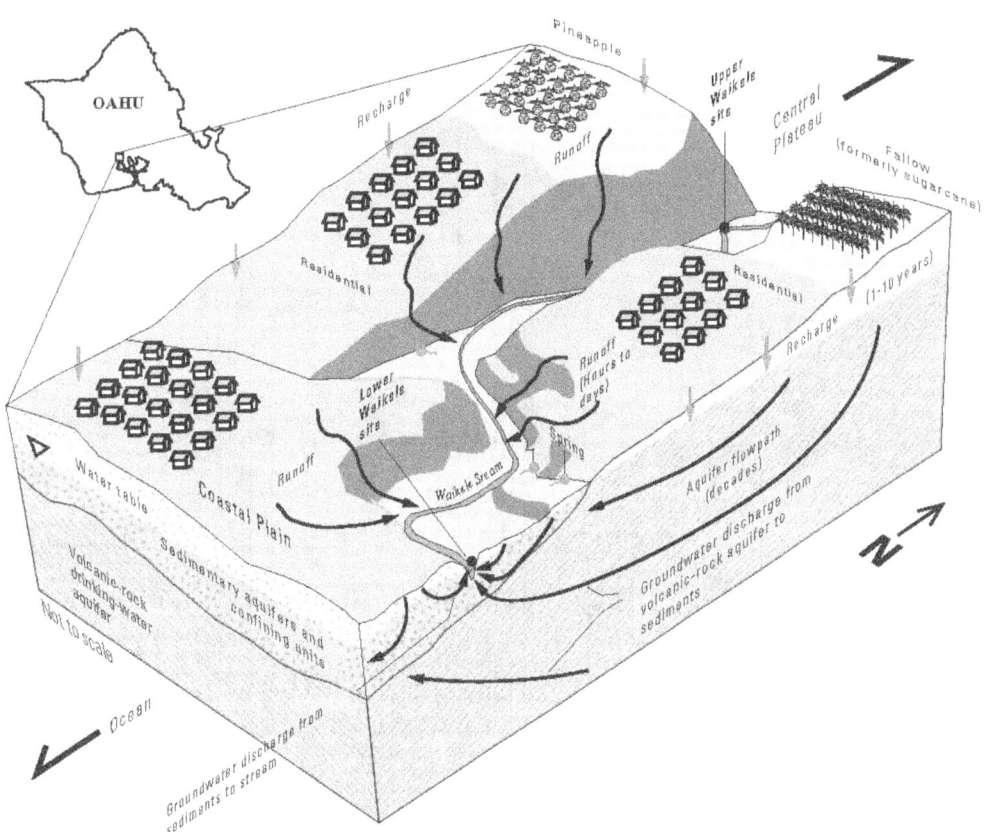

Figure 42. Block diagram showing hydrologic transport pathways on central Oahu. From Anthony and others (2004), modified from a block diagram provided by Scot Izuka.

Once water recharges the water table, it becomes part of the regional groundwater flow system within the aquifer or aquifer system. Flow is mostly lateral over great distances, but also downward to deeper aquifers or parts of the flow system in recharge-dominated areas and upward to points of discharge at the downgradient ends of the flow systems. Contaminants in the groundwater may be drawn into drinking-water wells along the way or may affect receiving waters in rivers or lakes, for example in cases where groundwater-borne nutrients contribute to eutrophication (excessive growth of aquatic plants or algal blooms). Estimates of water transit time through the entire regional flow system range from a few decades for Oahu (Hunt, 1996) to as much as 350 years for flow through the entire Eastern Snake River Plain flow system (Ackerman, 1995).

The principal basaltic-rock and volcanic-rock aquifers are essentially unconfined aquifers, without regionally extensive clay layers to inhibit downward infiltration. Rising water tables have been recorded in the Columbia Plateau

and Snake River Plain within days or weeks of seasonal filling of irrigation canals or induced recharge experiments (for example Wylie and others, 2008), and although some estimates of infiltration time to the deep water table on Oahu are on the order of years to a decade, much quicker (weeks or months) responses of the water table have been observed after heavy rains (fig. 43). These quick responses may be due in part to "fast pathways" through the volcanic rocks in the unsaturated zone, suggesting rapid infiltration of water from land surface through at least some part of the terrain. For example, the Schofield Shaft in the central Oahu plateau is an inclined tunnel excavated to the deep water table 600 feet below. Heavy rainfall in November 1954 ended a prolonged drought and was followed by several more months of heavy rain through March 1955. Despite the 600-foot depth to water at the site, the deep groundwater level began to rise 2–3 months after November and peaked 10 months later in September 1955 (fig. 43).

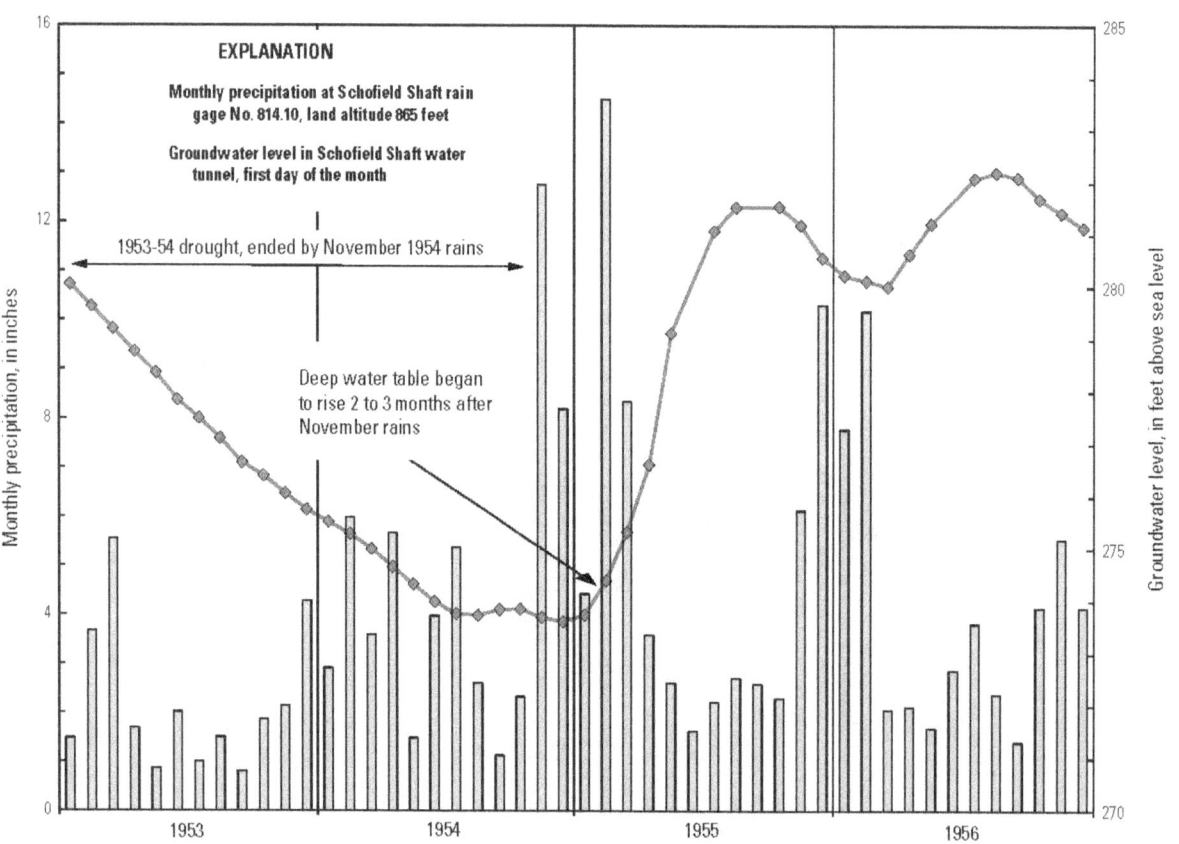

Figure 43.   Monthly precipitation and groundwater levels in the Schofield Shaft, central Oahu.

Irrigation

Irrigation has altered natural flow systems, both by diverting surface water from rivers and streams and by superimposing large artificial groundwater fluxes of "irrigation recharge." Because climate is semiarid in most cultivated areas of the Columbia Plateau, Snake River Plain, and Oahu, high crop productivity in all three areas is possible largely through the application of irrigation water diverted from rivers or pumped from underlying aquifers. Extensive irrigation canal systems divert water from the Columbia and Snake Rivers and distribute it across the landscape. Oahu has smaller scale irrigation ditches that divert water from wetter parts of the island.

When irrigation water is applied, not all of it is used by plants; much of it seeps beyond the reach of plant roots and infiltrates to deep groundwater, leaching dissolved fertilizer nutrients and other chemicals with it. This process is referred to in this report as "irrigation recharge" and is termed "irrigation-return recharge" in Hawaii. Other than direct seepage losses from canals and reservoirs, the amount of water infiltrating through irrigated fields depends on the irrigation method. "Furrow" or "field flood" irrigation is least efficient, losing the greatest fraction of applied water to deep infiltration. Estimates of furrow irrigation efficiency for sugarcane in Hawaii have centered around 50 percent, meaning that half the applied water is used by the plant (that is the "efficiency") and half is lost to deep infiltration. Spray irrigation is more efficient than flood, and drip irrigation through pinholes in plastic tubing is most efficient, estimated at 80–95 percent efficiency. Conversion to spray irrigation has progressed since the 1960s in the Columbia Plateau and Snake River Plain. Oahu converted from furrow to drip irrigation of sugarcane from the 1970s to mid-1980s (sugarcane cultivation ceased in 1996).

Although agricultural producers have transitioned to more efficient irrigation methods, prior decades of furrow/flood irrigation contributed large volumes of irrigation recharge, heavily augmenting the groundwater flow systems of the three study areas. Figure 44 illustrates the magnitude of irrigation-induced recharge in comparison to natural recharge under 1970s–1980s conditions. Mean annual recharge under natural conditions was estimated to be 5 inches or less over large expanses of non-irrigated rangeland in the Columbia Plateau and Snake River Plain. However, recharge exceeded 50 inches in some heavily irrigated areas along the Columbia and Snake Rivers (fig. 44). Data for Oahu show a similar pattern of higher than natural recharge in irrigated areas, as well as very high recharge (up to 214 inches) in the orographic belt of high rainfall along the east Oahu mountains.

Irrigation-induced recharge has raised groundwater levels by tens to hundreds of feet in the Columbia Plateau and Snake River Plain and has caused landslides from sedimentary bluffs (Phillips and others, 2008). On Oahu, irrigation has formed a distinct layer of degraded-quality water about 50–200 feet thick beneath the deep water table. This "irrigation-recharge layer" has been recognized since the 1960s and is discernible in geophysical borehole logs by higher water temperature and higher specific conductance than the naturally recharged groundwater beneath it. The high specific conductance corresponds to elevated concentrations of dissolved solids, such as nitrate, sulfate, bicarbonate, and other inorganic ions, and organic chemicals such as pesticides are also detected in the degraded-quality layer. Public-supply wells on Oahu typically are solid-cased for the first 100 feet or so below the water table in an attempt to exclude much of this degraded water from the wells.

Irrigation recharge has diminished in recent decades with conversion from flood irrigation to spray and drip irrigation. Recent droughts in the Columbia Plateau and Snake River Plain have decreased natural recharge and required additional pumping for irrigation. Groundwater levels have declined, raising concerns about long-term water availability. Drought also brings a water quality concern in the Snake River Plain, where quality may deteriorate because of less dilution by better quality natural recharge. On Oahu, irrigation recharge has diminished drastically, first with the conversion from furrow to drip irrigation of sugarcane by about 1980 and later when sugarcane cultivation ceased altogether in 1996. The poor-quality irrigation-recharge layer persists, however, and likely will take several decades or much of a century to be flushed from the aquifer.

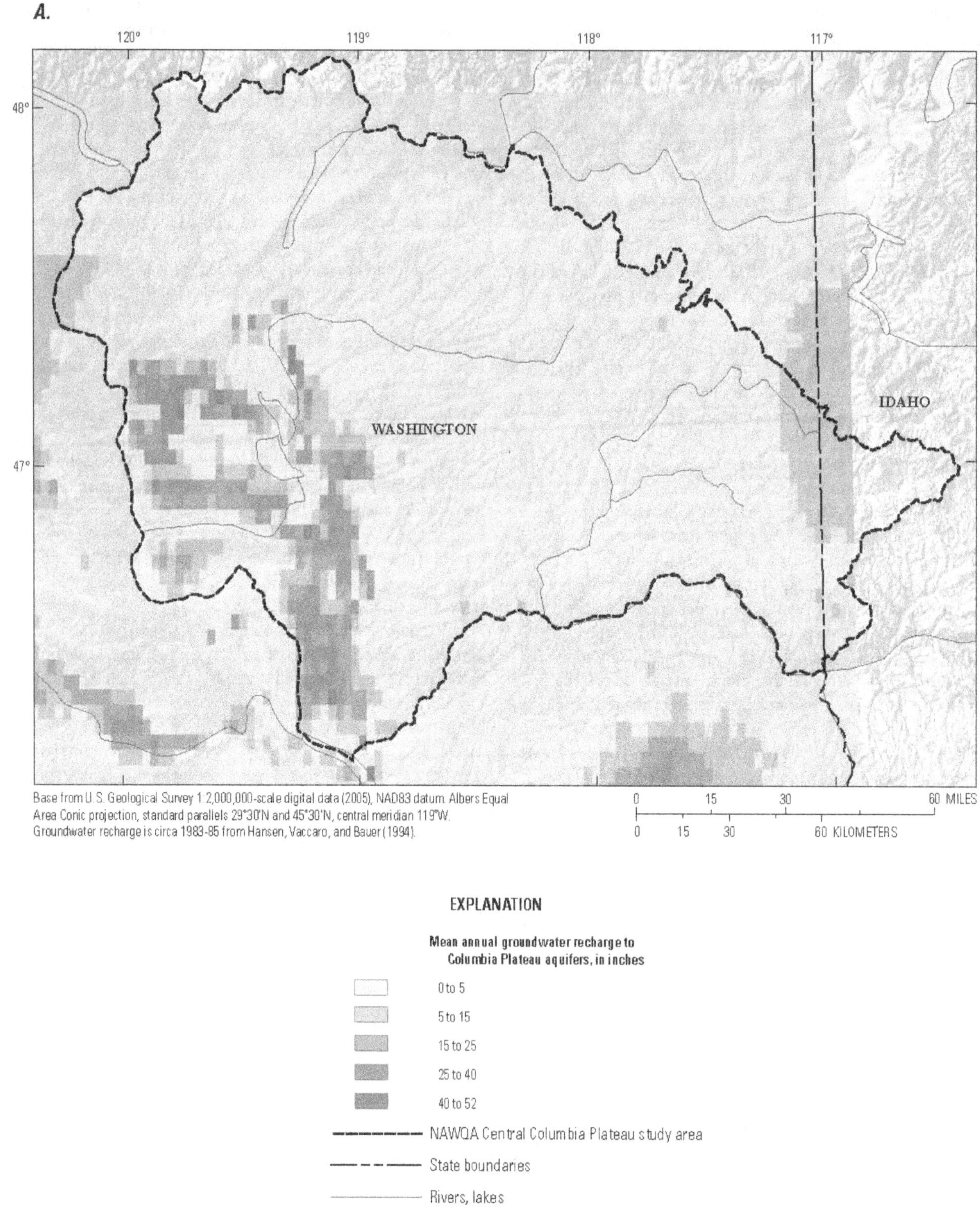

A.
120°
119°
118°
117°
48°
47°
WASHINGTON
IDAHO
Base from U.S. Geological Survey 1:2,000,000-scale digital data (2005), NAD83 datum. Albers Equal Area Conic projection, standard parallels 29°30'N and 45°30'N, central meridian 119°W.
Groundwater recharge is circa 1983-85 from Hansen, Vaccaro, and Bauer (1994).
0 15 30 60 MILES
0 15 30 60 KILOMETERS
EXPLANATION
Mean annual groundwater recharge to Columbia Plateau aquifers, in inches
0 to 5
5 to 15
15 to 25
25 to 40
40 to 52
NAWQA Central Columbia Plateau study area
State boundaries
Rivers, lakes

Figure 44. Mean annual groundwater recharge in (*A*) the Columbia Plateau, (*B*) the Eastern Snake River Plain, and (*C*) Oahu.

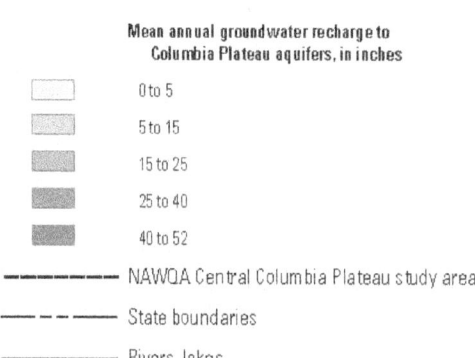

B.

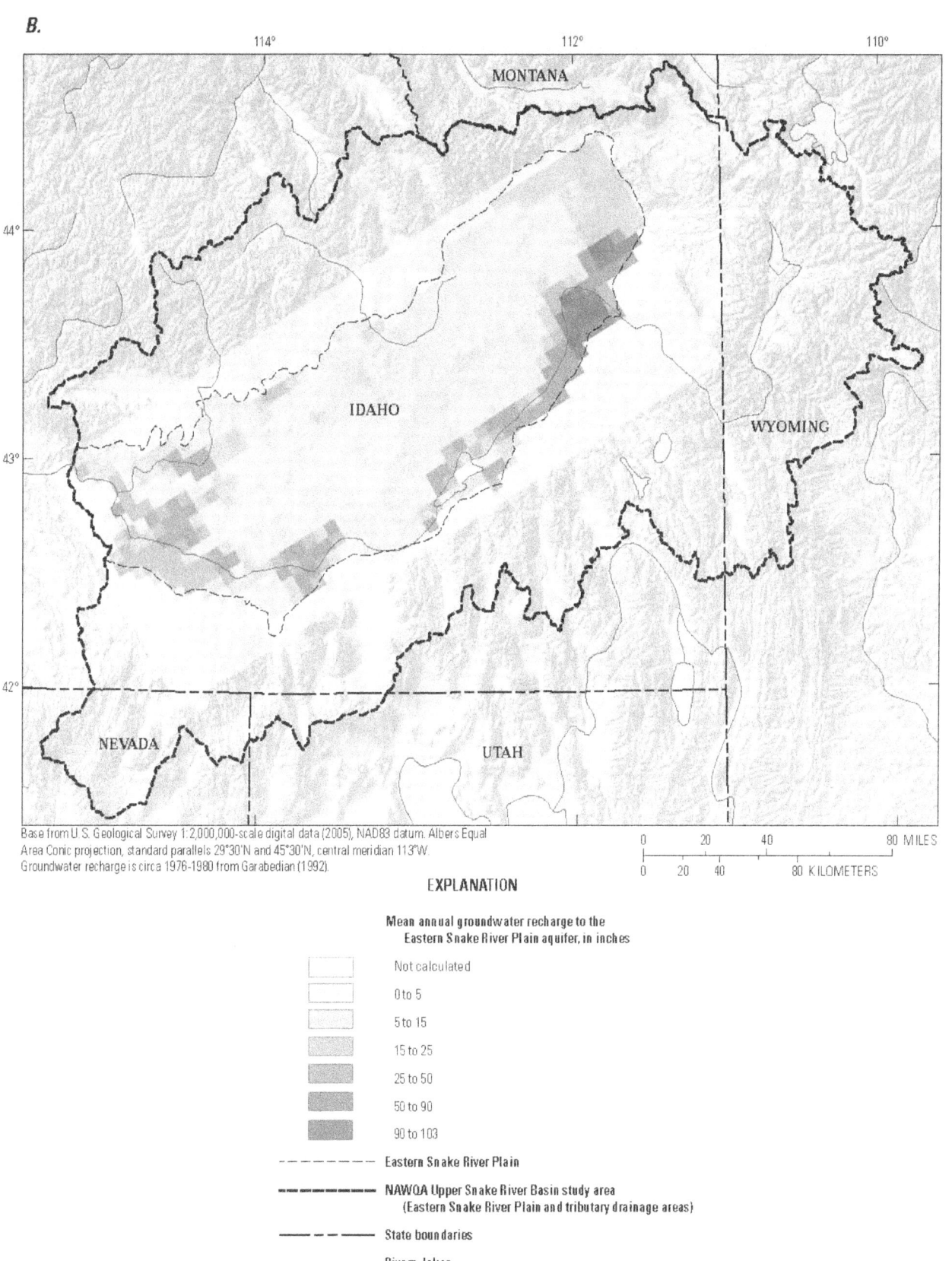

Base from U.S. Geological Survey 1:2,000,000-scale digital data (2005), NAD83 datum. Albers Equal
Area Conic projection, standard parallels 29°30'N and 45°30'N, central meridian 113°W.
Groundwater recharge is circa 1976-1980 from Garabedian (1992).

EXPLANATION

Mean annual groundwater recharge to the
Eastern Snake River Plain aquifer, in inches

Not calculated

0 to 5

5 to 15

15 to 25

25 to 50

50 to 90

90 to 103

– – – – – – Eastern Snake River Plain

━ ━ ━ ━ ━ NAWQA Upper Snake River Basin study area
(Eastern Snake River Plain and tributary drainage areas)

━ ━ ━ ━ ━ State boundaries

━━━━━━ Rivers, lakes

Figure 44.—Continued.

C.

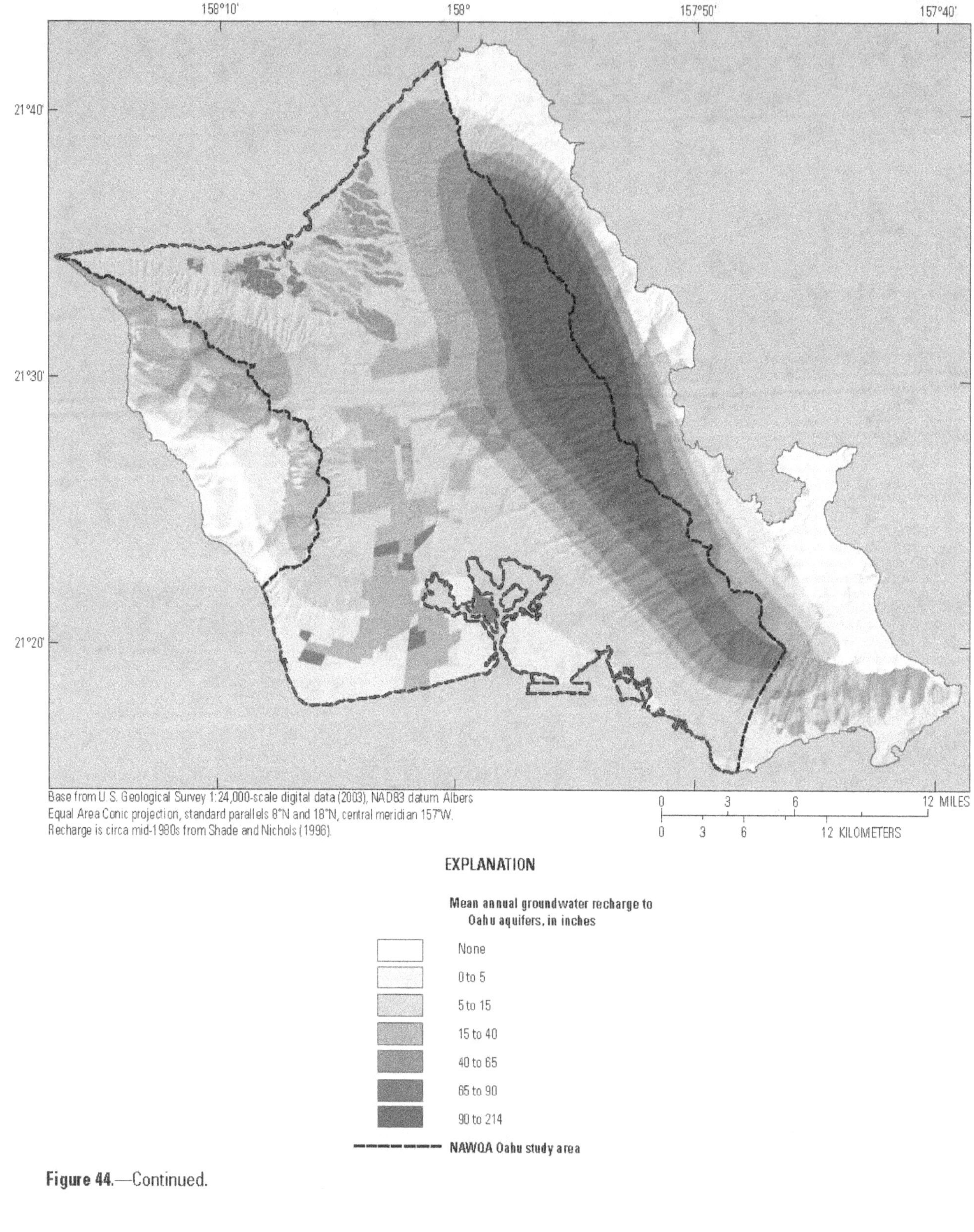

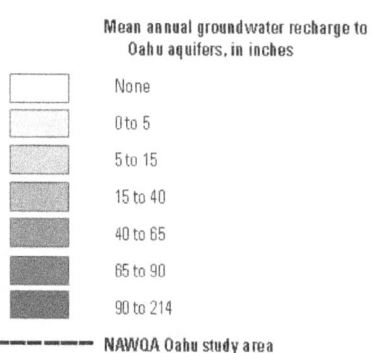

158°10'
158°
157°50'
157°40'
21°40'
21°30'
21°20'
Base from U.S. Geological Survey 1:24,000-scale digital data (2003), NAD83 datum Albers Equal Area Conic projection, standard parallels 8°N and 18°N, central meridian 157°W.
Recharge is circa mid-1980s from Shade and Nichols (1996).
0 3 6 12 MILES
0 3 6 12 KILOMETERS
EXPLANATION
Mean annual groundwater recharge to Oahu aquifers, in inches
None
0 to 5
5 to 15
15 to 40
40 to 65
65 to 90
90 to 214
NAWQA Oahu study area

Figure 44.—Continued.

Groundwater Recharge Dates

The recharge dates of many groundwater samples from the Columbia Plateau, Snake River Plain, and Oahu have been determined using several dating methods, including chlorofluorocarbons (CFCs), sulfur hexafluoride (SF6), tritium, and tritium-helium-3. CFCs and SF6 are manmade gases present at trace concentrations in the atmosphere and have a particular atmospheric concentration history that is known and can be matched to dissolved concentrations in groundwater to estimate the date that water was recharged (the time that water traveled through the unsaturated zone and reached the water table). CFCs were first developed to use as refrigerants in the early 1930s, so groundwater samples containing no CFCs are older than 1930 (Plummer and Busenberg, 1999). Significant production of SF6 began in the 1960s for use in high-voltage electrical switches (Busenberg and Plummer, 1997). Tritium in groundwater is the result of atmospheric testing of nuclear devices that began in 1952 and reached a maximum in 1963–64. Concentrations of tritium in precipitation have decreased since the mid-1960s peak of nuclear testing, except for some small increases from French and Chinese tests in the late 1970s (Solomon and Cook, 1999). Radioactive decay of tritium produces the noble gas helium-3. The ratio of tritium to helium-3 in groundwater is an effective dating tool for groundwaters younger than 1952 and extends the usefulness of tritium as an age-dating tracer up to the present day (2011).

Hinkle and others (2010) interpreted SF6 age data collected at 10 sites in the Columbia Plateau during 2002. The sampled wells were shallow monitoring wells (median depth to water 16.7 ft) completed in sedimentary deposits, not drawing from the regional basalt aquifer. The mean groundwater age was 17.5 years, and the median groundwater age was 16 years, indicating that the groundwater ages of the shallow waters in the sedimentary deposits are quite young.

Plummer and others (2000) collected CFC, tritium, and tritium-helium-3 samples from 48 wells and 5 springs in the Snake River Plain. They identified two types of groundwater: (1) regional background water that is unaffected by irrigation and fertilizer application and (2) mixtures of irrigation water from the Snake River and regional background water. Groundwater flow modeling by Garabedian (1992) and Ackerman (1995) indicates the age of the regional background water on the upgradient side of the Snake River Plain land-use studies to range between 50 and 150 years. The CFC, tritium, and tritium-helium-3 samples indicate that the young fraction of groundwater (that includes a mixture of irrigation water) is quite young. Samples from nearly 80 percent of the wells had ages younger than 10 years. Half of the samples were between 5 and 10 years. Only two samples had young fractions that were greater than 20 years old. The sources of the young fraction of groundwater are irrigation from surface water that leaches to the water table from the cultivated fields and leaky irrigation canals. Those irrigation waters also transport excess nitrate and anthropogenic chemicals to the water table (Plummer and others, 2000).

Apparent recharge dates of groundwater in Oahu using CFCs and SF6 were "young" for almost all samples, with only one date older than 1940 out of 45 wells (Hunt, 2004). The median groundwater recharge date for the 30 public-supply wells sampled in Oahu was 1966, and the median recharge date of the 15 monitoring wells was 1976. The samples were collected during 2000, so the median groundwater ages were 34 and 24 years, respectively. The young apparent ages and the prevalence of organic compounds in the same groundwater samples highlight the vulnerability of Hawaiian unconfined basalt aquifers to contamination: water recharges from land surface to the deep water table on a timescale of a few decades or less and carries anthropogenic chemicals with it.

Plots of nitrate concentrations in groundwater of the Columbia Plateau, Snake River Plain, and Oahu against groundwater recharge dates indicate that the increase of total fertilizer use in the United States is reflected in nitrate concentrations in groundwater (fig. 45). Before World War I, the primary sources of supplemental nitrogen for crops were animal manure, mineral sources such as potassium nitrate, and crop rotation with legume crops such as alfalfa. Synthetic fertilizers were first produced after World War I, when facilities that had produced ammonia and synthetic nitrates for explosives were converted to the production of N-based fertilizers (Rupert, 2008). Inorganic N fertilizer production was small until after World War II, when the production rates increased dramatically (fig. 45). Nationally, use of N fertilizer has increased rapidly from 1950 through about 1980 and then increased at a slower rate since about 1980. Nitrate concentrations in groundwater of the Columbia Plateau, Snake River Plain, and Oahu reflect that increase in fertilizer use, indicating that younger groundwaters have a higher probability of elevated nitrate concentrations.

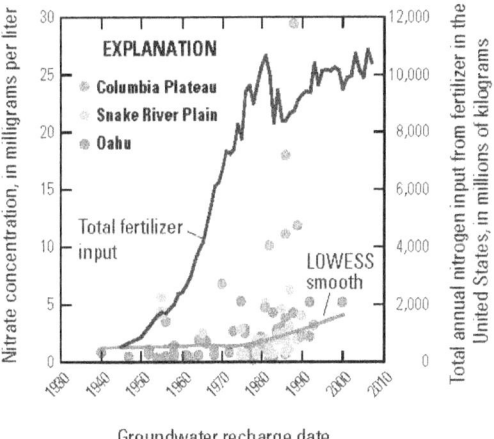

Figure 45. Nitrate concentrations and nitrogen input from fertilizer use versus groundwater recharge dates in the Columbia Plateau, Snake River Plain, and Oahu.

Land-Use Practices

Groundwater contamination reflects decades-old chemical releases and land-use practices in addition to recent conditions. Groundwater ages estimated from CFCs (chlorofluorocarbons) indicate that most groundwater sampled in the Columbia Plateau, Snake River Plain, and Oahu was recharged since 1940 or contains at least a fraction of young water recharged within that period. This period coincided with increasingly widespread and intensive manufacturing and use of chemicals in the second half of the 20th century. The presence of VOCs, pesticides, and elevated nitrate in many samples demonstrates that groundwater in the deep basaltic-rock and volcanic-rock aquifers is vulnerable to contamination from chemicals applied or released at land surface; water can travel from land surface to the deep water table within years and perhaps even months, carrying with it spilled or applied chemicals.

On the other hand, several compounds, such as EDB, DBCP, and DDT (whose breakdown product p,p'-DDE was detected) were discontinued from use in the late 1970s to early 1980s yet were still detected by NAWQA sampling decades later. The persistence of these "legacy" contaminants indicates that, although contamination may reach deep groundwater rapidly, it also takes decades for contaminants to flush out of the system.

Better chemical-use practices may be instituted from time to time, but groundwater systems are inherently slow in flushing or turnover time compared to streams and rivers. It can take decades to bring about a change in groundwater quality, and decades of strategic resampling are required to monitor and document those changes.

Redox Conditions

Groundwater samples indicated that redox conditions were predominantly in the oxic state in all of the sampled aquifers (fig. 46). Under oxic conditions, nitrate can persist for many years, even decades (McMahon and others, 2007). This was particularly true of the basaltic-rock and volcanic-rock aquifers, where samples were almost entirely oxic. Groundwater in unconsolidated sedimentary basin-fill aquifers of the Columbia Plateau and Snake River Plain also was predominantly oxic but slightly less so than the basaltic-rock aquifers.

The major-aquifer network in the Columbia Plateau basaltic-rock aquifer was an exception, in that a considerable number of samples indicated anoxic or mixed oxic-anoxic conditions. A mixed redox state of a groundwater sample is commonly thought to indicate that the well is drawing water from both oxic and anoxic zones or "microzones" in the aquifer, a reasonable supposition considering the sometimes strongly layered character of aquifer materials. Under reducing conditions, nitrate can degrade through denitrification, because the denitrifying bacteria prefer reducing conditions. Many pesticides such as atrazine degrade much faster under strongly reducing conditions than under oxic conditions. Other compounds, such as petroleum hydrocarbons, break down much faster under oxic conditions because the bacteria that consume petroleum hydrocarbons prefer oxic conditions.

Natural Processes

Some water-quality constituents originate from natural processes as water follows the groundwater hydrologic cycle from infiltration to groundwater flow and finally to points of groundwater discharge. Weathering and dissolution of rock minerals contributes some amount of dissolved mineral solids to groundwater, mostly major ions such as calcium, magnesium, sodium, bicarbonate, and sulfate. Plants also fix nitrogen from the atmosphere and derive phosphorus from soil minerals, imparting natural background concentrations of nitrate and phosphorus in water when the plants eventually die and decay and water percolates through the soil. Most naturally derived constituents cause little concern in the Western Volcanics aquifers, although arsenic is a notable exception. Arsenic is present at concentrations above the maximum contaminant level for drinking water in some Columbia Plateau (7 percent of wells sampled) and Snake River Plain drinking-water wells (3 percent of wells sampled), and although it may have leached from natural aquifer minerals, that leaching may have been fostered by infiltrating irrigation water (Busbee and others, 2009). Radon gas—a product of radioactive decay of uranium in the rocks—is also a natural constituent of concern for the Columbia Plateau and Snake River Plain.

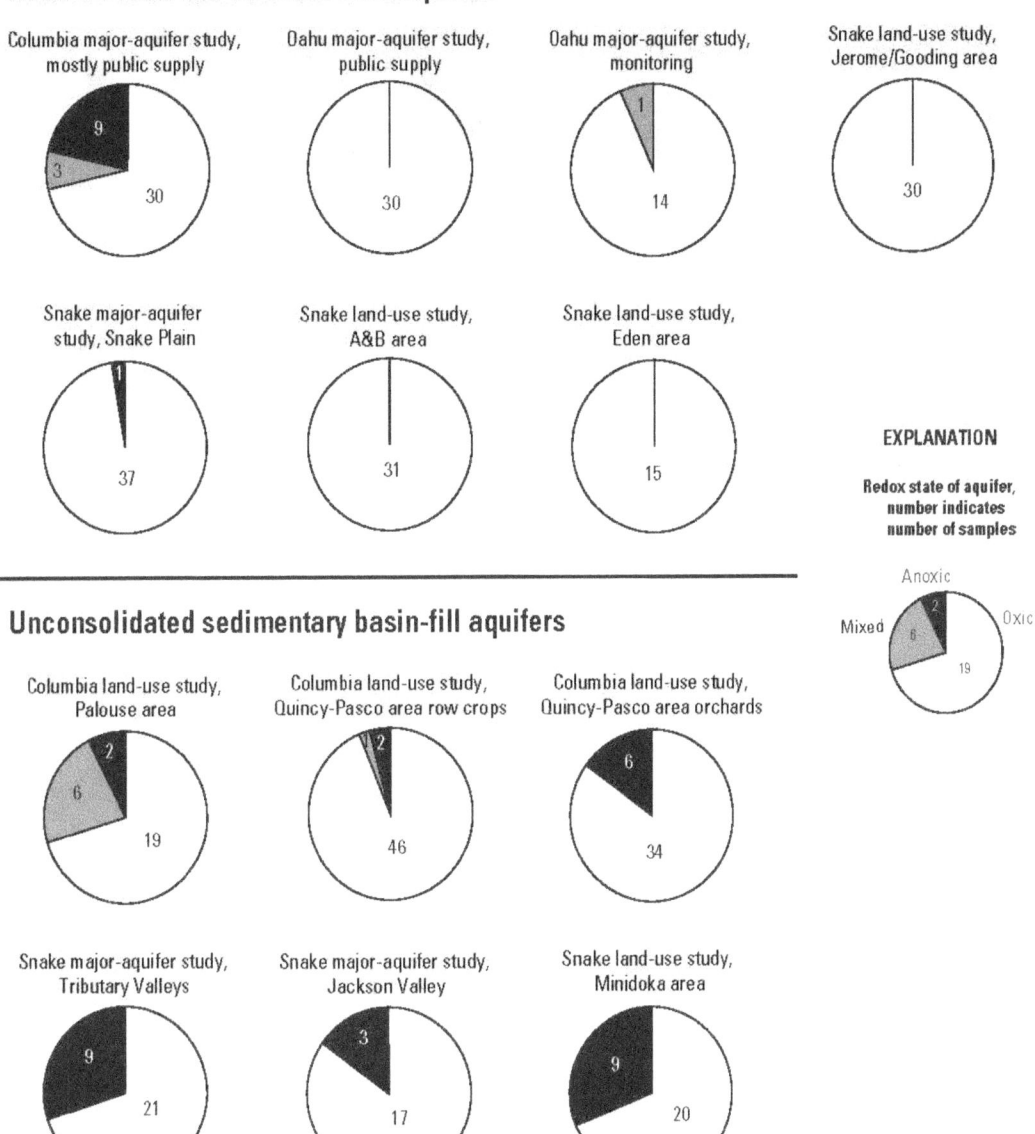

Figure 46. Redox conditions for groundwater sampled in wells in the Columbia Plateau, Snake River Plain, and Oahu.

Summary

This NAWQA assessment of groundwater-quality conditions of the Columbia Plateau, Snake River Plain, and Oahu for the period 1992–2010 shows where, when, why, and how specific water-quality conditions occur in groundwater of the three study areas and yields science-based implications for assessing and managing the quality of these water resources. The primary aquifers in the Columbia Plateau, Snake River Plain, and Oahu are mostly composed of fractured basalt, which makes their hydrology and geochemistry similar. In spite of the hydrogeologic similarities, there are climatic differences that affect the agricultural practices overlying the aquifers, which in turn affect the groundwater quality. Understanding groundwater-quality conditions and the natural and human factors that control groundwater quality is important because of the implications to human health, the sustainability of rural agricultural economies, and the substantial costs associated with land and water management, conservation, and regulation.

The principal regional aquifers of the Columbia Plateau, Snake River Plain, and Oahu are highly vulnerable to contamination by chemicals applied at the land surface; essentially, they are as vulnerable as many shallow surficial aquifers elsewhere. The permeable and largely unconfined character of principal aquifers in the Columbia Plateau, Snake River Plain, and Oahu allow water and chemicals to infiltrate to the water table despite depths to water commonly in the hundreds of feet. The aquifers are essentially unconfined over large areas, having few extensive clay layers to impede infiltration through permeable volcanic rock and alluvial sediments. Agriculture is intensive in all three study areas, and heavy irrigation has imposed large artificial flows of irrigation recharge, which rival or exceed natural recharge rates, over the natural flow systems. Fertilizers and pesticides applied at land surface are leached from soil and transported to deep water tables with the infiltrating irrigation recharge, resulting in a layer of degraded water quality overlying better quality regional groundwater beneath. This "irrigation-recharge layer" is best known on Oahu, where it has been studied since the 1960s; however the extent of nitrate and pesticide contamination in the Columbia Plateau and Snake River Plain indicate that the same situation exists there. Contamination from agricultural and urban activities is present not only at shallow depths in surficial materials of the three areas, but it extends over regional extent in the deep, principal bedrock aquifers that are tapped for drinking water by domestic and public-supply wells.

Both naturally occurring constituents and nitrate concentrations above human-health benchmarks were more common in the Columbia Plateau and the Snake River Plain; anthropogenic constituents (constituents related to human activities) above human-health benchmarks were more common in Oahu. Naturally occurring contaminants, such as arsenic and radon, may be present in groundwater at concentrations of potential concern for human health in relatively undeveloped settings that otherwise may not be perceived as susceptible to contamination. Even though the median depth to groundwater in Oahu is more than 300 feet, the common occurrence of anthropogenic compounds in groundwater indicates that Oahu has a high susceptibility to contamination.

Nitrate concentrations in groundwater were above the national background concentration of 1 milligram per liter (mg/L) in all three study areas. In the Columbia Plateau, nitrate exceeded the human-health benchmark of 10 mg/L in 17 percent of the wells sampled. In the Snake River Plain, nitrate exceeded the human-health benchmark of 10 mg/L in 3 percent of the wells sampled. Nitrate can persist in groundwater for years and even decades, so prudent groundwater protection measures are critical to protect drinking water resources.

Nitrate logistic regression models indicated that areas with a high percentage of land in crops (such as potatoes or sugarcane), and soils with low amounts of organic matter, are most likely to have elevated nitrate concentrations in the groundwater. Areas where agricultural activities were absent had much lower probabilities of detecting elevated nitrate concentration. The Columbia Plateau had a much higher probability of having elevated nitrate concentrations, with most of the land area having greater than a 50-percent probability of elevated nitrate concentrations. Oahu and the Snake River Plain had a much lower probability of having elevated nitrate concentrations because of their lower percentage of agricultural land.

Pesticides were detected frequently in groundwater of the Columbia Plateau, Snake River Plain, and Oahu, but generally at low concentrations below human-health benchmarks. Atrazine and its degradate (a compound produced from the breakdown of a parent pesticide), deethylatrazine, were the most commonly detected pesticides in groundwater sampled in the Columbia Plateau and Snake River Plain. Bromacil was the most commonly detected pesticide on Oahu. The other pesticides most commonly detected in the study areas include simazine, hexazinone, metribuzin, diuron, prometon, metolachlor, p,p'-DDE, dieldrin, 2-4-D, and alachlor. DDE (a degradate of DDT) and dieldrin are still being detected in groundwater because of historical use despite having been banned for more than 30 years. Codetection of multiple pesticides in water from a single well was common. The widespread occurrence of pesticides in groundwater in the study areas indicates that the groundwater is highly susceptible to pesticide contamination.

Some pesticides were detected in groundwater samples from all three study areas, but other pesticides were detected only in samples from Oahu or only in samples from the Columbia Plateau and Snake River Plain. This is because some pesticides (such as atrazine) are broad-spectrum pesticides that are used on many crops in many different areas of the United States. Other pesticides (such as simazine, metribuzin, and metolachlor) are used on row crops (such as potatoes, barley, and alfalfa) grown in the Columbia Plateau and Snake River Plain, but not on pineapple or sugarcane grown in Oahu.

Atrazine logistic regression models indicate that areas with a high percentage of land in crops (such as potatoes or sugarcane), a low percentage of fallow land, and highly permeable soils with low amounts of organic matter are most likely to have atrazine detected in the groundwater. Areas where agricultural activities were absent had much lower probabilities of atrazine being detected. The Snake River Plain had a much higher probability of atrazine detections, with more than 50 percent of the land area having greater than a 50-percent probability of atrazine contamination. Oahu had a much lower probability of atrazine contamination, with only 24 percent of the land area having greater than a 50-percent probability of atrazine contamination.

Oahu and the Columbia Plateau had some of the highest percentages of soil fumigant detections in groundwater anywhere in the United States. Soil fumigants are volatile organic compounds (VOCs) used as pesticides and are applied to soils to reduce populations of plant parasitic nematodes (harmful rootworms), weeds, fungal pathogens, and other soil-borne microorganisms. They are used in Oahu and the Columbia Plateau on crops such as pineapple and potatoes. All three areas (Columbia Plateau, Snake River Plain, and Oahu) had fumigant concentrations that exceed human-health benchmarks for drinking water.

References Cited

Ackerman, D.J., 1995, Analysis of steady-state flow and advective transport in the eastern Snake River Plain Aquifer System, Idaho: U.S. Geological Survey Water-Resources Investigations Report 94-4257, 25 p. (Also available at http://pubs.er.usgs.gov/publication/wri944257.)

Ackerman, Frank, 2007, The economics of atrazine: International Journal of Occupational and Environmental Health, v. 13, no. 4, p. 441–449, accessed March 26, 2012, at http://ase.tufts.edu/gdae/Pubs/rp/EconAtrazine.pdf.

Alexander, R.B., and Smith, R.A., 1990, County-level estimates of nitrogen and phosphorus fertilizer use in the United States, 1945 to 1985: U.S. Geological Survey Open-File Report 90-130, 12 p. (Also available at http://pubs.er.usgs.gov/publication/ofr90130.)

Anthony, S.S., Hunt, C.D., Jr., Brasher, A.M., Miller, L.D., and Tomlinson, M.S., 2004, Water quality on the island of Oahu, Hawaii, 1999–2001: U.S. Geological Survey Circular 1239, 32 p. (Also available at http://pubs.er.usgs.gov/publication/cir1239.)

Ayotte, J.D., Szabo, Zoltan, Focazio, M.J., and Eberts, S.M., 2011, Effects of human-induced alteration of groundwater flow on concentrations of naturally occurring trace elements at water-supply wells: Applied Geochemistry, v. 26, no. 5, p. 747–762, accessed March 27, 2012, at http://www.sciencedirect.com/science/article/pii/S088329271100045X.

Barbash, J.E., and Resek, E.A., 1996, Influence of pesticide properties, environmental setting, and study design on pesticide detections, chap. 6 of Pesticides in ground water—Distribution, trends, and governing factors: Chelsea, Mich., Ann Arbor Press, Inc., p. 269–311.

Bartsch, R., Forderkunz, S., Reuter, U., Sterzl-Eckert, H., and Griem, H., 1998, Maximum workplace concentration values and carcinogenicity classification for mixtures: Environmental Health Perspectives, v. 106, supp. 6, p. 1291–1293.

Battaglin, W.A., and Goolsby, D.A., 1995, Spatial data in geographic information system format on agricultural chemical use, land use, cropping practices in the United States: U.S. Geological Survey Water-Resources Investigations Report 94-4176, 87 p. (Also available at http://pubs.er.usgs.gov/publication/wri944176.)

Bender, D.A., Zogorski, J.S., Halde, M.J., and Rowe, B.L., 1999, Selection procedure and salient information for volatile organic compounds emphasized in the National Water-Quality Assessment Program: U.S. Geological Survey Open-File Report 99-182, 32 p. (Also available at http://pubs.er.usgs.gov/publication/ofr99182.)

Brasher, A.M., and Anthony, S.S., 2000, Occurrence of organochlorine pesticides in stream bed sediment and fish from selected streams on the island of Oahu, Hawaii, 1998: U.S. Geological Survey Fact Sheet 140-00, 6 p. (Also available at http://pubs.er.usgs.gov/publication/fs14000.)

Burlinson, N.E., Lee, L.A., and Rosenblatt, D.H., 1982, Kinetics and products of hydrolysis of 1,2-dibromo-3-chloropropane: Environmental Science and Technology, v. 16, no. 9, p. 627-632.

Busbee, M.W., Kocar, B.D., and Benner, S.G., 2009, Irrigation produces elevated arsenic in the underlying groundwater of a semi-arid basin in Southwestern Idaho: Applied Geochemistry, v. 24, no. 5, p. 843–859.

Busenberg, E., and Plummer, L.N., 1997, Use of sulfur hexafluoride as a dating tool and as a tracer of igneous and volcanic fluids in ground water [abs.]: Geological Society of America, Annual Meeting, Salt Lake City, 1997, Abstracts with Programs, v. 29, no. 6, p. A-78.

California Environmental Protection Agency, 2009, Public health goal for 1,2,3-trichloropropane in drinking water: California Environmental Protection Agency, Pesticide and Environmental Toxicology Branch, Office of Environmental Health Hazard Assessment, accessed March 28, 2012, at http://www.oehha.ca.gov/water/phg/pdf/082009TCP_phg.pdf.

Carpenter, D.O., Arcaro, K.F., Bush, B., Niemi, W.D., Pang, S., and Vakharia, D.D., 1998, Human health and chemical mixtures—An overview: Environmental Health Perspectives, v. 106, supp. 6, p. 1263–1270.

Childress, C.J.O., Foreman, W.T., Conner, B.F., and Maloney, T.J., 1999, New reporting procedures based on long-term method detection levels and some considerations for interpretations of water-quality data provided by the U.S. Geological Survey National Water Quality Laboratory: U.S. Geological Survey Open-File Report 99-193, 19 p. (Also available at http://pubs.er.usgs.gov/publication/ofr99193.)

Clark, G.M., Maret, T.R., Rupert, M.G., Maupin, M.A., Low, W.H., and Ott, D.S., 1998, Water quality in the upper Snake River Basin, Idaho and Wyoming, 1992–95: U.S. Geological Survey Circular 1160, 35 p. (Also available at http://pubs.er.usgs.gov/publication/cir1160.)

Clark, Heather, and Snedeker, Suzanne, 2004, Pesticides and breast cancer risk—Dibromochloropropane (DBCP): Ithaca, N.Y., Cornell University, Program on Breast Cancer and Environmental Risk Factors, Fact Sheet #50, July 2004, accessed March 28, 2012, at http://envirocancer.cornell.edu/FactSheet/pesticide/fs50.dbcp.cfm.

Daly, C., Neilson, R.P., and Phillips, D.L., 1994, A statistical-topographic model for mapping climatological precipitation over mountainous terrain: Journal of Applied Meteorology, v. 33, p. 140–158.

Daly, C., Taylor, G.H., and Gibson, W.P., 1997, The PRISM approach to mapping precipitation and temperature, in reprints: 10th Conference on Applied Climatology, Reno, Nev., American Meteorological Society, p. 10–12.

DeSimone, L.A., Hamilton, P.A., and Gilliom, R.J., 2009, The Quality of our Nation's waters—Quality of water from domestic wells in principal aquifers of the United States, 1991–2004—Overview of major findings: U.S. Geological Survey Circular 1332, 49 p. (Also available at http://pubs.er.usgs.gov/publication/cir1332.)

Dubrovsky, N.M., Burow, K.R., Clark, G.M., Gronberg, J.M., Hamilton P.A., Hitt, K.J., Mueller, D.K., Munn, M.D., Nolan, B.T., Puckett, L.J., Rupert, M.G., Short, T.M., Spahr, N.E., Sprague, L.A., and Wilber, W.G., 2010, The quality of our Nation's waters—Nutrients in the Nation's streams and groundwater, 1992–2004: U.S. Geological Survey Circular 1350, 174 p. (Also available at http://pubs.er.usgs.gov/publication/cir1350.)

Frans, L.M., 2000, Estimating the probability of elevated nitrate (NO_2+NO_3-N) concentrations in ground water in the Columbia Basin Ground Water Management Area, Washington: U.S. Geological Survey Water-Resources Investigations Report 2000-4110, 24 p. (Also available at http://pubs.er.usgs.gov/publication/wri004110.)

Frans, L.M., and Helsel, D.R., 2005, Evaluating regional trends in ground-water nitrate concentrations of the Columbia Basin Ground Water Management Area, Washington: U.S. Geological Survey Scientific Investigations Report 2005-5078, 14 p. (Also available at http://pubs.er.usgs.gov/publication/sir20055078.)

Garabedian, S.P., 1992, Hydrology and digital simulation of the regional aquifer system, eastern Snake River plain, Idaho: U.S. Geological Survey Professional Paper 1408-F, 102 p., 10 pls. (Also available at http://pubs.er.usgs.gov/publication/pp1408F.)

GeoLytics, 2001, Census 2000 Short Form Blocks, 2000 Census of Population and Housing: E. Brunswick, N.J., GeoLytics, Inc., CD-ROM.

Giambelluca, T.W., Nullet, M.A., and Schroeder, T.A., 1986, Rainfall atlas of Hawai'i: Honolulu, Department of Land and Natural Resources, Hawaii Division of Water and Land Development Report R76, 26 p., accessed March 28, 2012, at http://www.wrcc.dri.edu/pcpn/hawaiitext.pdf.

Gilliom, R.J., Alley, W.M., and Gurtz, M.E., 1995, Design of the National Water-Quality Assessment Program—Occurrence and distribution of water-quality conditions: U.S. Geological Survey Circular 1112, 33 p. (Also available at http://pubs.er.usgs.gov/publication/cir1112.)

Gilliom, R.J., Barbash, J.E., Crawford, C.G., Hamilton, P.A., Martin, J.D., Nakagaki, Naomi , Nowell, L.H., Scott, J.C., Stackelberg, P.E., Thelin, G.P., and Wolock, D.M., 2006, The quality of our Nation's waters—Pesticides in the Nation's streams and ground water, 1992–2001: U.S. Geological Survey Circular 1291, 173 p. (Also available at http://pubs.er.usgs.gov/publication/cir1291.)

Hansen, A.J., Vaccaro, J.J., and Bauer, H.H., 1994, Ground-water flow simulation of the Columbia Plateau regional aquifer system, Washington, Oregon, and Idaho: U.S. Geological Survey Water-Resources Investigations Report 91-4187, 81 p., 15 pls. (Also available at http://pubs.er.usgs.gov/publication/wri914187.)

Harding Lawson Associates, 1995, Final sampling and analysis plan for Operable Unit 2 Phase II remedial investigation, Schofield Army Barracks, Island of Oahu Hawaii: Prepared for U.S. Army Environmental Center, Aberdeen Proving Ground, Md., report and appendixes variously paged.

Helsel, D.R., and Hirsch, R.M., 1992, Statistical methods in water resources: New York, Elsevier Science Publishing Company, Inc., 522 p.

Hess, C.T., Michel, J., Horton, T.R., Prichard, H.M., and Coniglio, W.A., 1985, The occurrence of radioactivity in public water supplies in the United States: Health Physics, v. 48, no. 5, p. 553–586.

Hinkle, S.R., Shapiro, S.D., Plummer, L.N., Busenberg, Eurybiades, Widman, P.K., Casile, G.C., and Wayland, J.E., 2010, Estimates of tracer-based piston-flow ages of groundwater from selected sites—National Water-Quality Assessment Program, 1992–2005: U.S. Geological Survey Scientific Investigations Report 2010-5229, 90 p. (Also available at http://pubs.er.usgs.gov/publication/sir20105229.)

Hirsch, R.M., Alley, W.M., and Wilber, W.G., 1988, Concepts for a National Water-Quality Assessment Program: U.S. Geological Survey Circular 1021, 42 p.

Homer, C., Huang, C., Yang, L., Wylie, B., and Coan, M., 2004, Development of a 2001 national landcover database for the United States: Photogrammetric Engineering and Remote Sensing, v. 70, no. 7, p. 829–840.

Hopke, P.K., Borak, T.B., Doull, J., Cleaver, J.E., Eckerman, K.F., Gundersen, L.C.S., Harley, N.H., Hess, C.T., Kinner, N.E., Kopecky, K.J., McKone, T.E., Sextro, R.G., and Simon, S.L., 2000, Health risks due to radon in drinking water: Environmental Science and Technology, v. 34, no. 6, p. 921–926.

Hosmer, D.W., and Lemeshow, Stanley, 2000, Applied logistic regression: New York, John Wiley and Sons, Inc., 375 p.

Hunt, C.D., Jr., 1996, Geohydrology of the island of Oahu, Hawaii: U.S. Geological Survey Professional Paper 1412-B, 55 p. (Also available at http://pubs.er.usgs.gov/publication/pp1412B.)

Hunt, C.D., Jr., 2004, Ground-water quality and its relation to land use on Oahu, Hawaii, 2000-01: U.S. Geological Survey Water-Resources Investigations Report 2003-4305, 57 p. (Also available at http://pubs.er.usgs.gov/publication/wri20034305.)

Idaho Agricultural Statistics Service, 1999, Idaho Agricultural Statistics: Idaho Agricultural Statistics Service, 65 p.

Idaho Department of Environmental Quality, 2000, Cumulative impacts assessment, Thousand Springs area of the Eastern Snake River Plain, Idaho: Ground Water Quality Technical Report No. 14, 48 p.

Jones, J.L., and Wagner, R.J., 1995, Water-quality assessment of the central Columbia Plateau in Washington and Idaho; analysis of available nutrient and pesticide data for ground water, 1942–92: U.S. Geological Survey Water-Resources Investigations Report 94-4258, 119 p. (Also available at http://pubs.er.usgs.gov/publication/wri944258.)

Kahle, S.C., Morgan, D.S., Welch, W.B., Ely, D.M., Hinkle, S.R., Vaccaro, J.J., and Orzol, L.L., 2011, Hydrogeologic framework and hydrologic budget components of the Columbia Plateau Regional Aquifer System, Washington, Oregon, and Idaho: U.S. Geological Survey Scientific Investigations Report 2011–5124, 66 p. (Also available at http://pubs.er.usgs.gov/publication/sir20115124.)

Klasner, F.L., and Mikami, C.D., 2003, Land use on the island of Oahu, Hawaii, 1998: U.S. Geological Survey Water-Resources Investigations Report 2002-4301, 20 p. (Also available at http://pubs.er.usgs.gov/publication/wri20024301.)

Kleinbaum, D.G., 1994, Logistic regression, a self-learning text: New York, Springer-Verlag, 282 p.

Koterba, M.T., Wilde, F.D., and Lapham, W.W., 1995, Ground-water data-collection protocols and procedures for the National Water-Quality Assessment Program—Collection and documentation of water-quality samples and related data: U.S. Geological Survey Open-File Report 95-399, 114 p. (Also available at http://pubs.er.usgs.gov/publication/ofr95399.)

Larson, S.J., Gilliom, R.J., and Capel, P.D., 1999, Pesticides in streams of the United States—Initial results from the National Water-Quality Assessment Program: U.S. Geological Survey Water-Resources Investigations Report 98-4222, 92 p. (Also available at http://pubs.er.usgs.gov/publication/wri984222.)

Lindholm, G.F., 1996, Summary of the Snake River plain regional aquifer-system analysis in Idaho and eastern Oregon: U.S. Geological Survey Professional Paper 1408-A, 59 p., 1 pl. (Also available at http://pubs.er.usgs.gov/publication/pp1408A.)

Lindholm, G.F., and Vaccaro, J.J., 1988, Region 2—Columbia Lava Plateau, chap. 5 of Back, W., Rosenshein, J.S., and Seaber, P.R., eds., Hydrogeology—The geology of North America, v. 0-2: Boulder, Colo., Geological Society of America, p. 37-50.

Lindsey, B.D., and Rupert, M.G., 2012, Methods for evaluating temporal groundwater quality data and results of decadal-scale changes in chloride, dissolved solids, and nitrate concentrations in groundwater in the United States, 1988-2010: U.S. Geological Survey Scientific Investigations Report 2012-5049, 46 p. (Also available at http://pubs.er.usgs.gov/publication/sir20125049.)

Macdonald, G.A., Abbott, A.T., and Peterson, F.L., 1983, Volcanoes in the sea—The geology of Hawaii (2d. ed.): Honolulu, Hawaii, University of Hawaii Press, 517 p.

Maupin, M.A., and Barber, N.L., 2005, Estimated withdrawals from principal aquifers in the United States, 2000: U.S. Geological Survey Circular 1279, 52 p. (Also available at http://pubs.er.usgs.gov/publication/cir1279.)

McMahon, P.B., Dennehy, K.F., Bruce, B.W., Gurdak, J.J., and Qi, S.L., 2007, Water-quality assessment of the High Plains aquifer, 1999–2004: U.S. Geological Survey Professional Paper 1749, 136 p. (Also available at http://pubs.er.usgs.gov/publication/pp1749.)

Menard, Scott, 2002, Applied logistic regression analysis, second edition: Sage Publications Inc., 111 p.

Miller, J.A., ed., 1999, Ground water atlas of the United States: U.S. Geological Survey Hydrologic Atlas 730 A-H. (Also available at http://pubs.usgs.gov/ha/ha730/.)

Mundorff, M.J., Crosthwaite, E.G., and Kilburn, Chabot, 1964, Ground water for irrigation in the Snake River basin in Idaho: U.S. Geological Survey Water Supply Paper 1654, 224 p. (Also available at http://pubs.er.usgs.gov/publication/wsp1654.)

National Academy of Sciences, 1999, Health effects of exposure to radon: Washington, D.C., National Academy Press, 516 p.

National Research Council, 1999, Arsenic in drinking water: Washington, D.C., National Academy Press, 273 p.

Nelson, L.M., 1991, Surface-water resources for the Columbia Plateau in parts of Washington, Oregon, and Idaho: U.S. Geological Survey Water-Resources Investigations Report 88-4105, 4 pls. (Also available at http://pubs.er.usgs.gov/publication/wri884105.)

Nolan, B.T., and Clark, M.L., 1997, Selenium in irrigated agricultural areas of the western United States: Journal of Environmental Quality, v. 26, no. 3, p. 849–857.

Nolan, B.T., and Hitt, K.J., 2003, Nutrients in shallow ground waters beneath relatively undeveloped areas in the conterminous United States: U.S. Geological Survey Water-Resources Investigations Report 2002-4289, 17 p. (Also available at http://pubs.er.usgs.gov/publication/wri20024289.)

Nolan, B.T., Ruddy, B.C., Hitt, K.J., and Helsel, D.R., 1998, A National look at nitrate contamination of ground water: Water Conditioning and Purification, v. 39, no. 12, p. 76–79.

Oki, D.S., and Brasher, A.M.D., 2003, Environmental setting and the effects of natural and human-related factors on water quality and aquatic biota, Oahu, Hawaii: U.S. Geological Survey Water-Resources Investigations Report 2003-4156, 98 p. (Also available at http://pubs.er.usgs.gov/publication/wri20034156.)

Otton, J.K., 1992, The geology of radon: U.S. Geological Survey General Interest Publication, 29 p. (Also available at http://pubs.er.usgs.gov/publication/7000018.)

Pesticide Management Education Program, 1993, Pesticide Information Profile, Hexazinone: Ithaca, N.Y., Cornell University Web site, Pesticide Management Education Program, accessed March 30, 2012, at http://pmep.cce.cornell.edu/profiles/extoxnet/haloxyfop-methylparathion/hexazinone-ext.html.

Phillips, W., Garwood, D., and Stewart, R., 2008, Landslide hazards of Idaho: Idaho Geological Survey GeoNote G-44.

Plummer, L.N., and Busenberg, E., 1999, Chlorofluorocarbons—Tools for dating and tracing young groundwater, chap. 15 of Cook, P. and Herczeg, A., eds., Environmental tracers in subsurface hydrology: Boston, Kluwer Academic Publishers, p. 441–478.

Plummer, L.N., Busenberg, Eurybiades, and Cook, P.G., 2006, Principles of chlorofluorocarbon dating, in International Atomic Energy Agency, Use of chlorofluorocarbons in hydrology—A guidebook: Vienna, International Atomic Energy Agency, p. 17–29.

Plummer, L.N., Rupert, M.G., Busenberg, E., and Schlosser, P., 2000, Age of irrigation water in ground water from the Eastern Snake River Plain aquifer, south-central Idaho: Ground Water, v. 38, no. 2, p. 264–283, accessed March 29, 2012, at http://pubs.er.usgs.gov/publication/70022945.

Pratt, J.W., 1959, Remarks on zeros and ties in the Wilcoxon signed rank procedures: Journal of the American Statistical Association, v. 54, no. 287, p. 655–667, accessed March 29, 2012, at http://www.jstor.org/stable/2282543.

Price, C., Nakagaki, N., Hitt, K., and Clawges, R., 2003, Mining GIRAS—Improving on a national treasure of land use data, in Proceedings of the 23rd ESRI International Users Conference, July 7–11, 2003, Redlands, Calif., 11 p.

Ruddy, B.C., Lorenz, D.L., and Mueller, D.K., 2006, County-level estimates of nutrient inputs to the land surface of the conterminous United States, 1982–2001: U.S. Geological Survey Scientific Investigations Report 2006-5012, 17 p. (Also available at http://pubs.er.usgs.gov/publication/sir20065012.)

Rungvetvuthivitaya, M., Ray, C., and Green, R.E., 2007, A post-audit study of ground water contamination in the Pearl Harbor Aquifer: Journal of Hydrologic Engineering, v. 12, no. 6, p. 549–558.

Rupert, M.G., 1996, Major sources of nitrogen input and loss in the upper Snake River basin, Idaho and western Wyoming, 1990: U.S. Geological Survey Water-Resources Investigations Report 96-4008, 15 p. (Also available at http://pubs.er.usgs.gov/publication/wri964008.)

Rupert, M.G., 1997, Nitrate (NO_2+NO_3-N) in ground water of the upper Snake River Basin, Idaho and western Wyoming, 1991-95: U.S. Geological Survey Water-Resources Investigations Report 97-4174, 47 p. (Also available at http://pubs.er.usgs.gov/publication/wri974174.)

Rupert, M.G., 2008, Decadal-scale changes of nitrate in ground water of the United States, 1988–2004: Journal of Environmental Quality, v. 37, p. S-240–S-248.

Schwarz, G.E., and Alexander, R.B. 1995, State Soil Geographic (STATSGO) data base for the conterminous United States: U.S. Geological Survey Open-File Report 95-449. (Also available at http://pubs.er.usgs.gov/publication/ofr95449.)

Shade, P.J., and Nichols, W.D., 1996, Water budget and the effects of land-use changes on ground-water recharge, Oahu, Hawaii: U.S. Geological Survey Professional Paper 1412-C, 38 p.

Skinner, K.D., and Donato, M.M., 2003, Probability of detecting elevated concentrations of nitrate in ground water in a six-county area of south-central Idaho: U.S. Geological Survey Water-Resources Investigations Report 2003-4143, 29 p. (Also available at http://pubs.er.usgs.gov/publication/wri034143.)

Solomon, D.K., and Cook, P.G., 1999, ^{3}H and ^{3}He, in Cook, P.G., and Herczeg, A.L., eds., Environmental tracers in subsurface hydrology: Boston, Kluwer Academic Publishers, p. 397–424.

SPSS, Inc., 2000, SYSTAT 10, Statistics I—Software documentation: Chicago, SPSS, Inc., 663 p.

State of Hawaii, 2011, Amendment and compilation of chapter 11-20, Hawaii Administrative rules relating to potable water systems: State of Hawaii, Department of Health, no pagination, accessed May 2, 2012, at http://gen.doh.hawaii.gov/sites/har/AdmRules1/11-20.pdf.

SYSTAT Software, 2004, SYSTAT 11, Statistics II—Software documentation: Richmond, Calif., SYSTAT Software, Inc., 657 p.

Tenorio, P.A., Young, R.H.F., and Whitehead, H.C., 1969, Identification of return irrigation water in the subsurface—Water quality: Honolulu, University of Hawaii Water Resources Research Center Technical Report No. 33, 90 p.

Thomas, D.M., Paillet, F.L., and Conrad, M.E., 1996, Hydrogeology of the Hawaii scientific drilling project borehole KP-1-2—Groundwater geochemistry and regional flow patterns: Journal of Geophysical Research, v. 101, no. B5, p. 11683–11694.

Toccalino, P.L., Norman, J.E., Booth, N.L, Thompson, J.L., and Zogorski, J.S., 2012, Health-based screening levels: benchmarks for evaluating water-quality data: U.S. Geological Survey, National Water-Quality Assessment Program, accessed May 2, 2012, at http://water.usgs.gov/nawqa/HBSL/.

Tolan, T.L., Martin, B.S., Reidel, S.P., Kauffman, J.D., Garwood, D.L., and Anderson, J.L., 2009, Stratigraphy and tectonics of the central and eastern portions of the Columbia River Flood-Basalt Province—An overview of our current state of knowledge, in O'Connor, J.E., Dorsey, R.J., and Madin, I.P., eds., Volcanoes to vineyards—geologic field trips through the dynamic landscape of the Pacific Northwest: Geological Society of America Field Guide 15, p. 645–672.

U.S. Census Bureau, 2008, Population estimates and projections: U.S. Census Bureau, digital data accessed October 17, 2008, at http://www.census.gov/popest/.

U.S. Census Bureau, 1991, Census of population and housing, 1990—Public Law 94-171 data (United States): Washington, D.C., U.S. Census Bureau, [CD-ROM].

U.S. Department of Agriculture, 1991, State soil geographic data base (STATSGO): U.S. Department of Agriculture, Soil Conservation Service, Misc. Publication no. 1492, 88 p.

U.S. Department of Agriculture, 1993, Soil survey manual: U.S. Department of Agriculture, Handbook no. 18, 437 p.

U.S. Department of Agriculture, 1995, Soil survey geographic (SSURGO) data base: U.S. Department of Agriculture, Natural Resources Conservation Service, Misc. Publication no. 1527, 31 p.

U.S. Department of Agriculture, 2010, Quick Stats User Guide: U.S. Department of Agriculture, Natural Resources Conservation Service, accessed March 30, 2012, at http://quickstats.nass.usda.gov/help.

U.S. Department of Health and Human Services, 2002, DDT, DDE, and DDD: U.S. Department of Health and Human Services, Agency for Toxic Substances and Disease Registry, Division of Toxicology TaxFAQs, CAS # 50-29-3, 72-55-9, 72-54-8, 2 p., accessed March 29, 2012, at http://www.atsdr.cdc.gov/tfacts35.pdf.

U.S. Environmental Protection Agency, 1975, DDT—A review of scientific and economic aspects of the decision to ban its use as a pesticide: U.S. Environmental Protection Agency, EPA-540/1-75-022, 300 p., accessed March 29, 2012, at http://www.epa.gov/aboutepa/history/topics/ddt/DDT.pdf .

U.S. Environmental Protection Agency, 1983, EPA acts to ban EDB pesticide: EPA Press Release, September 30, 1983, accessed March 29, 2012, at http://www.epa.gov/aboutepa/history/topics/legal/02.html.

U.S. Environmental Protection Agency, 1995, R.E.D. facts, Metolachlor: U.S. Environmental Protection Agency, Office of Prevention, Pesticides, and Toxic Substances, EPA-738-F-95-007, 14 p., accessed March 29, 2012, at http://www.epa.gov/oppsrrd1/REDs/factsheets/0001fact.pdf.

U.S. Environmental Protections Agency, 1996, R.E.D. facts, Bromacil: U.S. Environmental Protection Agency, Office of Prevention, Pesticides, and Toxic Substances, EPA-738-F-96-013, 12 p., accessed March 30, 2012, at http://www.epa.gov/oppsrrd1/REDs/0041red.pdf.

U.S. Environmental Protection Agency, 2000, Final close-out report, Schofield Army Barracks, Oahu, Hawaii: Superfund Records Center 3315-00044, DD40, 18 p. plus appendixes.

U.S. Environmental Protection Agency, 2003a, Reregistration eligibility decision (RED) for diuron: U.S. Environmental Protection Agency, Office of Pesticide Programs, accessed March 30, 2012, at http://www.epa.gov/oppsrrd1/REDs/diuron_red.pdf.

U.S. Environmental Protection Agency, 2003b, Health effects support document for metribuzin: U.S. Environmental Protection Agency, Office of Water (4303T), EPA 822-R-03-004, 84 p., accessed March 30, 2012, at http://water.epa.gov/action/advisories/drinking/upload/2004_01_16_support_ccl_metribuzin_healtheffects.pdf.

U.S. Environmental Protection Agency, 2003c, Health effects support document for Aldrin/Dieldrin: U.S. Environmental Protection Agency, EPA 822-R-03-001, 255 p., accessed March 30, 2012, at http://water.epa.gov/action/advisories/drinking/upload/2004_1_16_support_ccl_aldrin-dieldrin_healtheffects.pdf.

U.S. Environmental Protection Agency, 2003d, Atrazine interim reregistration eligibility decision (IRED) Q&A's—January 2003: U.S. Environmental Protection Agency Web site, accessed March 30, 2012, at http://www.epa.gov/pesticides/factsheets/atrazine.htm.

U.S. Environmental Protection Agency, 2004, Pesticides industry sales and usage, 2000 and 2001 market estimates: U.S. Environmental Protection Agency, Office of Prevention, Pesticides, and Toxic Substances, EPA-733-R-04-001, accessed March 30, 2012, at http://www.epa.gov/opp00001/pestsales/01pestsales/market_estimates2001.pdf.

U.S. Environmental Protection Agency, 2005, Overview of the use and usage of soil fumigants: U.S. Environmental Protection Agency Memorandum, accessed March 22, 2011, at http://www.regulations.gov/fdmspublic/ContentViewer?objectId=09000064800b2f57&disposition=attachment&contentType=pdf.

U.S. Environmental Protection Agency, 2008, Health effects support document for 1,3-dichloropropene: Washington, D.C., U.S. Environmental Protection Agency, EPA Document Number: 822-R-08-008, January 2008, 142 p., accessed April 14, 2011, at http://www.epa.gov/ogwdw000/ccl/pdfs/reg_determine2/healtheffects_ccl2-reg2_13dichloropropene.pdf .

U.S. Environmental Protection Agency, 2009a, National Primary Drinking Water Regulations: U.S. Environmental Protection Agency, EPA 816-F-09-004, accessed March 30, 2012, at http://water.epa.gov/drink/contaminants/index.cfm#Primary.

U.S. Environmental Protection Agency, 2009b, A citizen's guide to radon—The guide to protecting yourself and your family from radon: U.S. Environmental Protection Agency, EPA 402/K-09/001, 16 p., accessed March 30, 2012, at http://www.epa.gov/radon/pdfs/citizensguide.pdf.

U.S. Environmental Protection Agency, 2011a, About pesticides—What is a pesticide: U.S. Environmental Protection Agency Web site, accessed March 30, 2012, at http://www.epa.gov/pesticides/about/.

U.S. Environmental Protection Agency, 2011b, Basic information about simazine in drinking water: U.S. Environmental Protection Agency Web site, accessed June 27, 2011, at http://water.epa.gov/drink/contaminants/basicinformation/simazine.cfm.

U.S. Environmental Protection Agency, 2011c, Atrazine updates: U.S. Environmental Protection Agency Web site, accessed March 30, 2012, at http://www.epa.gov/oppsrrd1/reregistration/atrazine/atrazine_update.htm.

U.S. Environmental Protection Agency, 2011d, Proposed radon in drinking water regulation: U.S. Environmental Protection Agency Web site, accessed March 30, 2012, at http://water.epa.gov/lawsregs/rulesregs/sdwa/radon/regulations.cfm#Proposed%20Regulations.

U.S. Geological Survey, 1999, The quality of our Nation's waters—Nutrients and pesticides: U.S. Geological Survey Circular 1225, 82 p. (Also available at http://pubs.er.usgs.gov/publication/cir1225.)

U.S. Geological Survey, 2001, Selected findings and current perspectives on urban and agricultural water quality by the National Water-Quality Assessment Program: U.S. Geological Survey Fact Sheet 047-01, 2 p. (Also available at http://pubs.er.usgs.gov/publication/fs04701.)

Visher, F.N., and Mink, J.F., 1964, Ground-water resources in southern Oahu, Hawaii: U.S. Geological Survey Water-Supply Paper 1778, 133 p. (Also available at http://pubs.er.usgs.gov/publication/wsp1778.)

Vogelmann, J.E., Howard, S.M., Yang, L., Larson, C.R., Wylie, B.K., and Van Driel, N., 2001, Completion of the 1990s national land cover data set for the conterminous United States from Landsat Thematic Mapper data and ancillary data sources: Photogrammetric Engineering and Remote Sensing, v. 67, no. 6, p. 650–652.

Wanty, R.B., Lawrence, E.P., and Gundersen, L.C.S., 1992, A theoretical model for the flux of radon-222 from rock to ground water, in Gates, A., and Gunderssen, L.C.S., eds., Geologic controls on radon: Geological Society of America Special Paper 271, p. 73–78.

Wanty, R.B., and Nordstrom, D.K., 1993, Natural radionuclides, in Alley, W.W., ed., Regional Ground-Water Quality: New York, Van Nostrand Reinhold, p. 423–441.

Ward, M.H., Mark, S.D., Cantor, K.P., Weisenburger, D.D., Correa-Villaseñor, A., and Zahm, S.H., 1996, Drinking water nitrate and the risk of non-Hodgkin's lymphoma: Epidemiology, v. 7, p. 465–471.

Welch, A.H., Watkins, S.A., Helsel, D.R., and Focazio, M.J., 2000, Arsenic in ground-water resources of the United States: U.S. Geological Survey Fact Sheet 063-00, 4 p. (Also available at http://pubs.er.usgs.gov/publication/fs06300.)

Weyer, P.J., Cerhan, J.R., Kross, B.C., Hallberg, G.R., Kantamneni, J., Breuer, G., Jones, M.P., Zheng, W., and Lynch, C.F., 2001, Municipal drinking water nitrate level and cancer risk in older women—The Iowa women's health study: Epidemiology, v. 12, no. 3, p. 327–338.

Whitehead, R.L., 1992, Geohydrologic framework of the Snake River Plain regional aquifer system, Idaho and eastern Oregon: U.S. Geological Survey Professional Paper 1408-B, 32 p., 6 pls. (Also available at http://pubs.er.usgs.gov/publication/pp1408B.)

Whiteman, K.J., 1986, Ground-water levels in three basalt hydrologic units underlying the Columbia Plateau, Washington and Oregon, spring 1984: U.S. Geological Survey Water-Resources Investigations Report 86-4046, 4 pls. (Also available at http://pubs.er.usgs.gov/publication/wri864046.)

Williamson, A.K., Munn, M.D., Ryker, S.J., Wagner, R.J., Ebbert, J.C., and Vanderpool, A.M., 1998, Water quality in the central Columbia Plateau, Washington and Idaho, 1992–95: U.S. Geological Survey Circular 1144, 35 p. (Also available at http://pubs.er.usgs.gov/publication/cir1144.)

Wisconsin Department of Health Services, 2010, Human health hazards—Molybdenum in drinking water: Wisconsin Department of Health Services, Division of Public Health, P-00150, 2 p., accessed March 30, 2012, at http://www.dhs.wisconsin.gov/publications/P0/P00150.pdf.

Wolock, D.M., 1997, STATSGO soil characteristics for the conterminous United States: U.S. Geological Survey Open-File Report 97-656. (Also available at http://pubs.er.usgs.gov/publication/ofr97656.)

World Health Organization, 2011, Molybdenum in drinking water—Background document for development of WHO Guidelines for drinking-water quality: Geneva, Switzerland, World Health Organization, accessed March 30, 2012, at http://www.who.int/water_sanitation_health/dwq/chemicals/molybdenum.pdf.

Wylie, Allan, Quinn, Bill, McVay, Michael, Scheidt, Nicholas, and Pennington, Larry, 2008, Analysis of the 2007 post season recharge using North Side Canal: Boise, Idaho Department of Water Resources Open-File Report, 22 p.

Zhu, Y., and Li, Q.X., 2002, Movement of bromacil and hexazinone in soils of Hawaiian pineapple fields: Chemosphere, v. 49, no. 6, p. 671–676, accessed March 30, 2012, at http://www.sciencedirect.com/science/article/pii/S0045653502003922.

Zogorski, J.S., Carter, J.M., Ivahnenko, T., Lapham, W.W., Moran, M.J., Rowe, B.L., Squillace, P.J., and Toccalino, P.L, 2006, Volatile organic compounds in the nation's ground water and drinking-water supply wells: U.S. Geological Survey Circular 1292, 101 p. (Also available at http://pubs.er.usgs.gov/publication/cir1292.)

www.ingramcontent.com/pod-product-compliance
Lightning Source LLC
Chambersburg PA
CBHW081551170526
45166CB00009B/2660